VODKA

ROBERT VON GOEBEN

BILL MILNE

With a Foreword by Maurice Kanbar
Founder of Skyy Vodka

FRIEDMAN/FAIRFAX
PUBLISHERS

FRIEDMAN/FAIRFAX

PUBLISHERS

All photographs © Bill Milne 1999

Library of Congress Cataloging-in-Publication Data available upon request.

ISBN 1-56799-754-6

Editor: Ann Kirby
Art Director: Jeff Batzli
Designer: Galen Smith
Photography Director: Christopher Bain
Production Manager: Ingrid McNamara

Color separations by Colourscan Overseas Co. Pte Limited
Printed in Singapore by KHL Printing Co Pte Ltd

For bulk purchases and special sales, please contact:

Friedman/Fairfax Publishers
Attention: Sales Department
15 West 26th Street
New York, NY 10010
212/685-6610 FAX 212/685-1307

Visit our website:
www.metrobooks.com

AUTHOR DEDICATION

To my old man,

who finally found the wisdom to know the difference.

AUTHOR ACKNOWLEDGMENTS

How would I ever get a book like this done without the cheer-leading, prodding, editorializing, critiquing, psychoanalysis, and love of someone like Kathryn? Beats me.

Hats off to (in no particular order): Maurice Kanbar, David Kanbar, Kira Klaus, and Melissa Lilly at Skyy Spirits; Samara Farber at Kratz & Company; Michel Rouxz at Carillon Importers; Tito Beveridge at Fifth Generation; Michael Avitable at Rainbow Spirits; Owe Linner and Charlotta Haage at Absolut; Magnus Philipson at Znaps Vodka; Adam Sass at Paige Poulos; Christine Comeau at Kittling Ridge; Jim Columbo for the leads; Rob Robinson and Eric Sothern at Infusion; Rabah Abusbaltan at One Market Restaurant; Kristina and Bradley Rotter and Wivianne Runske for the many toasts; Peter Nordstrom for his Swedishness; Alex Reyfman for his Russian-ness; Andy Kieffer and Stina Kindwall for making it through the taste test; my mom, Sue, Doug, Billy, and Mack for the vodka meal (even if I didn't get the recipe); my mom again for always being in my corner; Maureen McCrane for the keen eye, Sherry Manis for another keen eye; Ann Kirby, a writer's dream of an editor; and finally, Spencer Weisbroth, my lawyer, who has neither fear nor loathing.

CONTENTS

FOREWORD

It may have all started at the beginning of human history, when a famished caveman went hunting for a deer or rabbit for dinner. He didn't catch an animal, but he found some rotting fruit on the ground—which in fact had started fermenting into alcohol—and decided to eat that in desperation. He discovered that it not only tasted pretty good, but it made him feel fantastic. He was ready to fight lions!

Reports today show that drinking alcohol in moderation can reduce stress. Stress kills, and a cocktail at the end of the day is one of the most civilized ways to relax and enjoy good company. I believe vodka is the finest way to consume alcohol. It is clean and colorless, so it tastes great in a martini, on the rocks, mixed in a classic cocktail, or infused with lively fruit flavors.

Vodka also contains a lower level of congeners than any other type of alcohol. Why is a low congener content important? Some years ago after experiencing a pounding headache from drinking a single cognac, I decided to study the causes of alcohol-related headaches. I found that they were due to congeners—the natural impurities that form during the fermentation of any carbohydrate into alcohol. I became determined to develop the cleanest vodka possible. After four years, I developed a unique four-step distillation and triple-filtration process, and introduced Skyy as the ultra-premium, low-congener vodka.

Today, vodka is an excellent quality beverage made by many skilled manufacturers. That's one reason why vodka is the most popular spirit in the world, and justifies this comprehensive guide. This book provides an extensive historical perspective of vodka. Robert Von Goeben's clever telling of the vodka story combined with Bill Milne's superb photography make it an essential reference for any vodka connoisseur, and I feel honored to have been asked to write this foreword.

Maurice Kanbar
Founder and Chairman, Skyy Spirits
San Francisco, California

ABOVE: *Skyy vodka founder Maurice Kanbar reflects on the cleanest of spirits at Skyy's San Francisco headquarters.*
OPPOSITE: *The perfect martini: ice cold and free of impurities.*

INTRODUCTION

I n the past, you'd saunter into a tavern more confident than your cohorts that when you reached the bar your choices were simple. You'd hang your coat on the rack, dust off the stool, and prepare to order. A bartender in a crisp white apron would have appeared before you and asked the inevitable question:

"What'll it be?"

This moment of decision was what separated you from your comrades. While they pondered over the myriad of single-malt scotches, the proliferation of premium wines, the plethora of exotic gins, your order paled in comparison. You simply said, "Vodka on the rocks."

IF YOU'RE A VODKA DRINKER, YOUR DAY HAS ARRIVED

Your pal behind the counter rarely hesitated. He'd grab a short glass, fill it with ice and a portion of equally clear spirits, and ask you to pony up the tab. By the time you had enjoyed your first sip, your friends were still deciding what continent their drinks would come from. You sat by and watched, smugly

thinking how uncomplicated your life was. Deep down, you jealously wished that your drink had the respect that brought with it choice, but you sipped your nameless vodka and kept your mouth shut. Who needs choices anyway?

Then one day you offered your obligatory order and the bartender hesitated. He peered over his shoulder at the increasing number of clear bottles behind him and asked, "Uh, what kind?" You froze, realizing that you'd advanced into the same league as your scotch-drinking friends. You finally had a choice besides that one anonymous bottle behind the counter.

Vodka has long been the Rodney Dangerfield of spirits. Considered by most people to be the tasteless liquor that got you drunk, vodka was the stuff that needed fruit to make it palatable. Or, more accurately, it was the fuel that gave a kick to otherwise impotent juices. It was the "bang" in a Harvey Wallbanger, the "drive" in a Screwdriver. Vodka, for a long time, was the faceless supporting cast member in a string of cheesy stage plays. If there was a drink that somehow needed a little alcoholic boost, vodka was the answer.

Somewhere along the way, the word got out that all kinds of vodka taste the same. Given the history of vodka making, this isn't surprising. Technically, vodka is made from fermenting "something." Anything. Early vodka-like concoctions came from fermenting different grains, potatoes, or even beets. In fact, when the Bureau of Alcohol, Tobacco, and Firearms came up with "Standards of Identity" for vodka, it didn't really do much to clarify vodka's nature—it merely stipulated that vodka should be distilled in such a way as to be "without distinctive character, aroma, taste, or color."

A faceless vodka might have been alright twenty years ago, but we've thankfully moved into a more discerning era. Wine achieved a permanent place on the top shelf, and no one was surprised when beer got a similarly upscale treatment with the proliferation of microbreweries during the past ten years. Then the martini craze hit in the early nineties, and a whole legion of upscale gin entered the market. Scotch also made the leap from a lowly soda companion to the guest star at exclusive tastings. And as a companion to this coming-out party for premium spirits, cigars finally made the long trek from the nickel smoke to a high-priced luxury for connoisseurs.

All the while, vodka waited on the sidelines. Until now.

Quality distillers from Stolichnaya to Smirnoff now indulge the faithful with a long list of ultrapremium vodkas. For those of you who have always enjoyed the lucent cleanness of vodka, it's now your turn to relish in the craftsmanship and attention previously reserved for other spirits. One crystal-clear glass of vodka is no longer just like another.

And for those of you who've enjoyed the addition of fresh fruit flavor to vodka, it's your turn as well. We've discarded the tutti-frutti blender drinks and replaced them with a more sophisticated concoction of flavors: the infusion. Just the sight of fresh fruits and exotic vegetables deliciously floating in ultrapremium vodka is enough to convince you that we've made the leap from quantity to quality. Believe me, we all breathed a sigh of relief when the bartender stopped suggesting a pint-size Screwdriver and started pouring precious portions from a decanter of high-end vodka, fresh fruit floating delicately within.

Given that, our vodka loyalist now bounds into the bar with all the vigor of the bourbon drinker and all the curiosity of the wine connoisseur. Today's vodka-lover is faced with not one but perhaps a half dozen exclusive varieties to choose from, and all are made with the distiller's newfound obsession with quality. The race to make the world's finest vodka has begun.

So the next time the bartender in the white apron asks, "What'll it be?" be ready for a different scene. More than likely, you might pause, relish the newfound attention, discard the stock answer, and proudly ask, "What vodkas have you got?"

PAGES 10 and 11: *Vodka has a rich heritage of distinctive packaging. Pictured are antique bottles from the private collection at the Levize distillery in St. Petersburg, Russia.*
OPPOSITE: *The Cold War may be over, but the hammer and sickle still hang in the distilleries of modern Russia.*

CHAPTER 1

"After the gruel, the ladies

of the house often went to

the medicine cabinet, and

there, washing down the vile

tasting gruel with vodka,

little by little got drunk and

emerged as assorted viragos,

lunatics, hysterics, and

finally flaming drunks."

Jedrzej Kitowicz

THE MURKY HISTORY OF THE CLEAR ELIXIR

THE FACT THAT DIFFERENT COUNTRIES CLAIM RIGHTS TO THE BIRTH OF VODKA IS NOT SURPRISING. THE LACK OF A "DEFINING EVENT" IN THE ORIGINS OF VODKA HAS MADE IT VIRTUALLY IMPOSSIBLE TO PINPOINT A PARTICULAR PLACE AND TIME OF ITS DISCOVERY. AND SINCE SO MANY DIFFERENT PEOPLE HAVE CONTRIBUTED TO VODKA'S EVOLUTION, IT'S ALSO TOUGH TO PIN A FOUNDER'S MEDAL SQUARELY ON THE CHEST OF ANY ONE PERSON. BUT THAT'S NOT TO SAY THAT HISTORIANS AND MARKETERS HAVEN'T TRIED.

Given the fact that premium vodka is now front and center in today's swaggering cocktail culture, there's no shortage of information and opinions on the development of this neutral spirit. This much is certain: even if this historical mystery isn't as clear as a glass of triple-distilled vodka, it's at least as much fun.

RUSSIA

First, let's state the obvious: no country is more identified with vodka than Russia. Much more than a popular drink, the "green dragon"—a Russian folk idiom for vodka—is an integral part of the culture. Vodka is at the heart of Russian society, the substance of both celebration and disheartenment. Yet though Russia is universally identified as ground zero for vodka, this still doesn't mean we can easily nail down the origin of the spirit.

Ask Russian historian William Pokhlebkin when vodka was invented, and be prepared to order another round before you get to the answer. Known as one of the leading historians on the subject, Pokhlebkin takes ninety-seven pages in his defining work, *A History of Vodka,* to reach the conclusion that "A date between 1448 and 1478 is supported by the whole sum of historical, economic, and social research," and that "the development of alcohol distilling . . . occurred in Russia approximately a century earlier than in neighboring countries."

As it turns out, alcohol was being produced in Russia long before it became the vodka we know today. Originally produced for medicinal purposes, these drinks (cloudy, contaminated, and very labor intensive) were more akin to crude wines than clear vodka. Early fermented liquids were known as *perevara,* a derivative of mead and beer, or *korchma,* meaning hooch or pot wine. They were made from any of a variety of sources: honey, grapes, even tree sap. Before the fifteenth century, an abundance of raw material, specifically honey, meant that there was plenty of mead, ale, and wine to go around.

ISIDOR AND THE GREAT ESCAPE:
Russian Monks and the Development of Vodka

IN THE FIFTEENTH CENTURY, monks were the most educated and technically proficient people in Russian society. It has been argued that the only way the religious leaders of the day could ever have sanctioned the production of alcohol was if it occurred within their own walls. But for the monks, alcohol distillation was not only possible, it was also quite profitable. There is no doubt that the Church and the Grand Prince of the Russian state were quite chummy back then. As a result of its comfortable position with the powers that be, the Church was one of the few entities exempt from the "tsar's duty" and other alcohol-related taxation. Folklore tells the story of one monk who used this close relationship with the state to save his own skin, and possibly discovered vodka in the process.

One of the brighter ecclesiastic minds of the day, Isidor was a member of the Russian delegation that ventured to Italy in the 1430s to attend the Eighth Ecumenical Council. It's been said that Isidor, while there, witnessed the art of distillation for the first time. It's also been said that he made the mistake of siding with the Roman pope, not the kind of move that went down well with the powers back home.

Returning to Russia, Isidor could easily have found himself a candidate for capital punishment, but instead he was held in the monastery of Chudov, under what we know today as "country club" conditions. According to legend, Isidor used the loose confines of his incarceration to practice the art of distillation that he had learned in Italy. And while the product of his still might not have been top shelf, it was most likely inebriating. Isidor supposedly used his spirits to lull the guards to sleep and escape. He ultimately made his way back to Rome, leaving behind the know-how that eventually produced today's vodka.

ABOVE: *Early vodkas may not have been the superpremium quality we enjoy today, but the bottles were top shelf.*
OPPOSITE: *The Archangel Cathedral, Moscow.*

By the fifteenth century, however, honey had become less available and there emerged a need for a cheaper raw material for alcohol. Grain was the answer, and its emergence may have come about as an offshoot of baking. Pokhlebkin says this could be the reason why no one type of grain is used for the production of distilled spirits. Many times, it emanated from the wastes of bread making, or from whatever else—potatoes, grain, molasses—was around.

Eventually the slow shift from wine and beer to more concentrated alcohol began to occur in Russia. When you think about it, this medieval transition to more concentrated spirits makes sense. The raw materials (mainly grain) were cheaper, the high concentration made the alcohol easier and

less expensive to transport, and it didn't spoil. The only thing that was needed was a steady supply of grain. The development of the three-field system of crop rotation in the 1440s provided just such a bounty. The resulting surplus meant that the needs for food were more frequently met, and thus more grain became available for the production of alcohol.

But to say that distillation occurred in a vacuum in Russia would be misleading. A couple of noteworthy visits to the Russian state in the fifteenth century certainly influenced the art of distillation. A Genoese delegation en route to Lithuania in 1426 presented *aqua vitae* to the Grand Prince of Russia, and demonstrated a technique for distillation. In the 1430s, a Russian church delegation visited Italy for the Eighth Ecumenical Council where members saw the production of alcohol firsthand.

As the technique for alcohol distillation became known in Russia, two trends emerged. First, due to its economic prosperity and abundant rye grain surpluses, Moscow became the de facto center of quality alcohol production. In fact, this capital city was so identified with the spirit that the term "Russian vodka" became synonymous with excellence. Muscovites touted their pure rye vodka, made from the soft "living" waters of the Moscow and Neva rivers, as the best in the world.

More importantly, fifteenth-century Moscow provided a strong, centralized state that made possible the second and most prevalent trend in the history of Russian vodka: the state monopoly. Over the next four hundred years, right up to and after the 1917 Russian revolution, state control over the production and sale of vodka became a monumental factor in both the development of governmental power and the often dicey relationship between Russia's aristocracy and her rulers.

Soon after vodka's emergence, Russian rulers saw it as a perfect vehicle for control and taxation. It generated enormous state revenues during the uncertain times of war, and taverns provided a convenient revenue funnel. Ivan the Terrible established the first *kabak* ("tsar's tavern") for members of his inner circle, the palace elite known as the *oprichnina*, in 1533. By the late seventeenth century, the tsar's taverns were so numerous that a visitor to Russia observed that they outnumbered bath houses.

However, this state stronghold also created another tradition: the right of the aristocratic class to distill their own alcohol. This noble privilege made sense given the endless supply of cheap labor and the ready availability of raw materials, mainly grain. In 1648, the first "tavern revolt" occurred in Moscow, caused by the inability of the poor to pay their tavern debts. This resulted in another period of state control.

The establishment of a strong state monopoly angered the aristocracy, encouraged widespread illicit production, and was not as profitable for the state. It also did little to develop the technology of distillation. These problems led to an establishment of mixed state-private systems. While this pacified the upper classes, who made great strides in technology, it caused widespread corruption and drunkenness among the populace (as well as a shrinkage in the state coffers), which eventually caused the state to invoke once again a monopoly.

This alternation between periods of state control and periods of privatization resulted in a seesaw of power that lasted until the eighteenth century, when control landed firmly in the hands of the privileged classes. But the aristocracy's hold on the vodka market was not all that long-lived. Through the following centuries, the government began to take more and more control, and by the time World War I erupted, the government was running the stills once again.

By the eighteenth century, distillation and filtration techniques were firmly in the hands of the upper classes, who had improved them to the point where Russian vodka started to resemble the spirit we know today. In that time, no one had a deeper appreciation for vodka than Tsar Peter the Great, who was said to order his vodka triple distilled, mixed with anise water, and distilled again. He was so enamored with his country's vodka that even the finest French liquors would not suffice. In 1717, on a trip to France, Peter wrote to his wife and complained about having to drink wine and cognac. "There is only one bottle of vodka left." he lamented, "I don't know what to do."

MOSCOW MULE

INGREDIENTS

2 parts vodka

1 part lime juice

Ginger beer

RECIPE

Pour vodka and lime juice over ice in a copper mug. Fill mug with ginger beer, and stir.

By the time of the Bolshevik revolution in 1917, widespread drunkenness had wreaked havoc on the working classes. As a result, one of the first acts of the new workers' government was to outlaw vodka. From the beginning of the revolution until the mid-1930s, the sale of vodka with an alcoholic strength of more than twenty percent was strictly prohibited. Lenin, Stephen White writes in his book *Russia Goes Dry*, commented that what the proletariat needed was "clarity, clarity, and once again clarity."

By most accounts, these strict curbs on vodka were effective. There emerged a public disdain for drunkenness, and a British delegation to Russia in 1924 commented on the noticeable absence of "the universal Russian drunkenness."

However, by 1936 Joseph Stalin had realized the potency of vodka as a tool for social control, and he soon began to loosen its restrictions. By the late 1930s, the alcohol content of vodka was allowed to increase to forty percent (its "natural" level). The cork finally came off the bottle during World War II when the Soviet Union began issuing vodka rations to soldiers. Massive consumption quickly became part of military culture. After the war, vodka prices continued to remain low and consumption soared. Vodka became a cheap anesthesia against the harsh realities of life in the USSR and remained so until the next attempts at controls in the 1980s.

POLAND

Although vodka is most closely associated with Russia, other countries stake a claim to its heritage as well. None participates in this battle of the bloodlines more vigorously than Poland. Some in this eastern European

OPPOSITE: *An antique etched glass vodka bottle from the Levize collection. For centuries, control of vodka production in Russia was caught in a struggle for power between the state, who relished the income from taxation, and the aristocracy, who demanded the right to run their own "estate" distilleries.*

ABOVE: *Rows of hydroponically cultivated peppers await harvesting for Canada's Inferno vodka.* **OPPOSITE:** *No doubt such flavors— and methods—were unheard of when this vintage Russian bottle was first filled with the white spirit.*

country even take credit for introducing vodka to Russia via the Baltic Republics. The Poles call the clear spirit *wodka,* a diminutive of *woda* (the Polish word for "water"), and claim that the name for their "little water" was ultimately borrowed by the Russians. (See the sidebar on page 54 for more on the vodka name war.)

The Polish tradition of alcohol production dates back to the twelfth century. Although predominately made from rye grain, early Polish spirits were also derived from herbs and fruits, starting this country's long tradition of flavored alcohol. These early beverages became known as *gorzale wino* ("burnt wine").

"To be healthy in Poland, pharmacies

and doctors should not be used,

but rather twice a year one should get

properly drunk."

Doctor's statement to Polish aristocrats,

late nineteenth century

Many of these concoctions were originally designed for medicinal purposes. The first encyclopedia of medicine and science, edited by Stefan Falimirz in 1534, included a chapter on distilling herb vodkas. It mentions vodka as being used for "washing the chin after shaving" and describes a "very delightful fragrant vodka which is rubbed on after washing in the bath, since it removes unpleasant odor from the person." Alcoholic elixirs became so identified as medical treatments that the name of one overzealous Lublin pharmacist, Michal Sztok, evolved into a common Polish expression for drunkenness.

Polish King Jan Olbracht in 1546 gave every Pole the right to produce and sell alcohol. The practice grew like wildfire throughout Poland until 1572, when the nobility lobbied for exclusive control. The result was an aristocratic privilege that put the right to alcohol production under the control of the gentry. This was a major windfall for the upper classes and the

IRON SKILLET ROASTED MUSSELS WITH LIME-CUMIN VODKA AND CHIPOTLE BUTTER

INGREDIENTS

Fresh mussels, washed and debearded

Sea salt

Pepper

½ pound butter

2 whole chipotle chiles, cut in half lengthwise

4 limes

1 liter Vodka

2 tablespoons whole cumin

RECIPE

To make chipotle butter: Toast the chiles lightly and add to melted butter. Simmer together on low heat for about 5 minutes. When done, add juice of the lime and salt to taste.

To make lime-cumin vodka: Add 3 of the limes, peeled and sectioned, and 2 tablespoons whole cumin to 1 liter vodka. Allow to sit at room temperature for 5 to 7 days.

Heat the skillet until hot, then add the mussels. (If you do not have a skillet, the mussels can be roasted open in a sauté pan.) Cover until the mussels have all opened. Sprinkle with sea salt and pepper and lightly douse with lime-cumin vodka. Serve with the melted chipotle butter.

Recipe courtesy of Infusion Bar & Restaurant, San Francisco, CA

PAGE 30: *While many consumers think of Russia when they think of vodka, it has become a truly global commodity in recent years. At Znaps distillery in the United Kingdom, vodka is made with ingredients imported from Sweden, and bottled on state-of-the-art equipment.* **PAGE 31:** *Though vodka comes out crystal clear, many enthusiasts prefer a bit of flavor—and color—in their vodka cocktails.*

country's coffers, which benefited from an accompanying tax. By 1580 Kraków, Gdansk, and Poznan became major centers for production, Poznan having over forty-nine working stills. Vodka was even used as a form of currency in some quarters.

Vodka was a national favorite by the seventeenth century, and carrying spirits in a small hip flask became very popular with aristocrats. Every local gentry had its own still, and production techniques steadily improved. Triple distillation began in the eighteenth century, yielding a purer and more potent vodka. Most varieties continued to be made from grain, but in the eighteenth century Poland introduced a raw material that would become synonymous with Polish vodka: the potato.

Potatoes were first grown in Poland at the palace of King Jan III Sobieski in Wilanow, the result of a gift from the emperor of Austria. Widespread cultivation began on royal estates in the eighteenth century and caught on rapidly as Poland's unique soil conditions (specifically along the Vistula River) were perfectly suited for the vegetable. Today, Poland is the world's second largest producer of potatoes. In fact, one of the country's most well-known potato vodka brands, Luksusowa, is named after the Polish word for "deluxe."

Polish distilleries continued to multiply under private ownership into the twentieth century, when a state monopoly was instituted in 1919. After World War II, this state monopoly was given the name Polmos, which consolidated control over the country's most well-known brands. Polmos, together with the state export monopoly, Agros, was privatized after the 1989 democratic elections in Poland, and still holds majority stakes in the country's most popular brands, including Wyborowa and Zubrowka.

SWEDEN

When vodka lovers think of Sweden, they usually think of one thing: Absolut. While the brand with the artistic ads has done a lot to propagate the notion of Sweden as a leading vodka producer, the history of this spirit starts long before Andy Warhol painted his first Absolut advertisement.

Swedish distillation, not surprisingly, given the fact that Sweden was not an independent country until 1905, emerged much like it did in neighboring countries. Spirits in Sweden were historically known as *brännvin* or *brännwein* ("burnt wines"). Even today, many Swedes use the phrase "to burn" when describing distillation.

Originally, Swedes had many uses for spirits outside of getting soused. Early distillates were used both as medicines and as an ingredient for making gunpowder. A Stockholm ledger dating from the 1460s contains an entry for "brännvin to make gunpowder." Sweden has a long history of tight controls over alcohol, and it wasn't until the end of the fifteenth century that the manufacture and sale of brännwein for anything other than gunpowder was legal. One fifteenth-century Swedish document claimed that brännwein could "cure more than 40 ailments, including headache, head lice, kidney stones, and toothache."

Home production had become a way of life in Sweden until 1756, when some 180,000 stills were confiscated in a nationwide sweep. But while home distillation was steadily curtailed (it was finally abolished in 1860), Swedes continued to exhibit a voracious appetite for vodkas of all flavors. Authors Nicholas Faith and Ian Wisniewski, in their book *Classic Vodka*, tell of nineteenth-century restaurants serving

ABOVE: *The private chef at a Siberian distillery. Trade secrets are closely guarded in the vodka industry. Most distilleries operate under heavy security, and some are watched by armed guards. Not surprisingly, the workers tend to eat their lunch in.*

canteens with up to six taps, offering "several different iced vodkas, including specially purified spirit and others blended with orange and caraway seeds, to be drunk with smörgåsbord." This appetite still continues today, as evidenced by Swedish vodka packaging. Rather than screw-off tops, vodka bottles in Sweden typically feature tear-off caps since a bottle is usually finished in one sitting.

The Swedish temperance movement that occurred in the nineteenth century (and the establishment of the state monopoly, Vin & Sprit, after World War II) placed the control of vodka squarely in the hands of the state, where it rests to this day. Outside the country's borders, Sweden was never really known for its vodka production until the 1980s. When the state-owned V&S Vin & Sprit AB connected with a New York advertising agency, the partnership marketed a simple vodka in an oddly shaped bottle, and Absolut became a household name in the United States.

OTHER NATIONALITIES

While Russia, Poland, and Sweden receive the lion's share of attention in the vodka world, noteworthy contenders exist around the globe. They may not be able to wave their national flags in the contest to claim the origins of vodka but they are still able to field some heavyweight contenders in taste and quality.

THE LANCUT DISTILLERY MUSEUM

TODAY'S VODKA VARIETIES ARE flush with historical legacies. But nowhere is vodka's rich cultural heritage more evident than in the southeastern region of Poland on the fringe of the Carpathian foothills. Here, along an old medieval trade route lies Lancut, Poland's oldest working distillery and home of the Distillery Museum.

The majestic Lancut Distillery and Museum, an imposing castle of stone ramparts, was originally the property of Stanislaw Stadnicki, known as the "Devil of Lancut." Stadnicki and his sons took part in a long history of war and misadventure and by 1628 the original Lancut castle was destroyed.

Under the control of the Lubomirska family, in the late eighteenth century pentagonal fortifications were constructed as a defense against the crusading Turks. These defenses took on a spectacular neoclassical flair, with exquisite stucco work designed by Italian artist Giovanni Battista Falconi. But in such a war-ravaged age, the castle's true beauty lies in its strength. It was considered impenetrable, and was able to contain eighty cannons and four hundred soldiers.

Art from all across Europe poured into Lancut thanks to the extravagant tastes of Princess Elzbieta Lubomirska. The royal family of Potocki took ownership of Lancut in 1830 and continued development on a grand scale, including the installation of an absurdly large number of luxurious baths, causing one visitor to compare it to Dianabad, the largest public bath in Vienna.

German troops used Lancut as an outpost during World War II. A week before the Soviets rolled into Lancut, Alfred III Potocki hightailed it to Vienna, but not before taking with him six hundred crates of priceless artwork.

Its artistic and military history notwithstanding, the distillery at Lancut continued to develop through the centuries. Today, the Lancut Distillery produces about ten percent of Poland's alcoholic spirits, including such well-known names as Wyborowa, Zubrowka, and Luksusowa.

ABOVE: *The Absolut bottle, inspired by an antique medicine bottle, has become one of the most recognized liquor bottles in the world.*
OPPOSITE: *Talk about having a cold one! Vodka is the spirit of choice 100 miles (160km) north of the Arctic circle at The Ice Hotel in Jukkasjärvi, Sweden. The arctic accommodations, sponsored by Absolut, melt in the spring, but thankfully theyrise like an icy phoenix in October.*

FINLAND

The most notable contender is Finland, which, like Sweden, touts a national brand that's as clean and crisp as its climate. The art of distillation is said to have come to Finland from mercenary soldiers returning home from a tour of duty in southern Europe. Distillation was largely a homespun industry until the construction of a high-capacity yeast plant in Rajamaki in 1888, which spawned a growing vodka business in Finland that has been slowed only by prohibition and the world wars.

The Finnish Rajamaki plant was pressed into service in World War II for the production of "Molotov cocktails." These alcohol-filled makeshift bombs were used against Soviet tanks after the Soviets' 1939 invasion of Finland. Since then, the Finlandia brand, with its distinctive iced bottle, has become known throughout the world as a top-shelf vodka.

THE UKRAINE, BALTIC REPUBLICS

The Ukraine has enjoyed a modicum of success in the vodka business since the breakup of the Soviet Union, mainly due to a distinctive rye vodka called *Gorilka* (after the Ukrainian word for vodka, meaning "burning"). The Baltic republics (Latvia, Lithuania, and Estonia) continue a long tradition in vodka production from their days as part of the Soviet Union. Most noteworthy are the Estonian brands Eesti Viin and Viru Valge.

VODKA SORBET IN NEST OF CANDY FLOSS

SORBET INGREDIENTS

¼ cup glucose

¾ cup sugar

⅔ cup lemon juice

⅓ cup vodka

RECIPE FOR SORBET

Boil glucose with sugar in 2½ cups water for approximately 1 minute. Add lemon juice.

Freeze in an ice cream machine or put the mixture in a bowl in the freezer, and stir with a whip every 20 minutes for about 3 hours.

CANDY FLOSS INGREDIENTS

¼ cup light corn syrup

2½ cups sugar

⅔ cup lemon juice

⅓ cup vodka

RECIPE FOR CANDY FLOSS

Mix sugar and corn syrup together and boil in ⅔ cup water until 142°C (275°F). Let the sugar mixture run in a very thin string over 2 cooking spoons to create a nest shape. Serve sorbet inside "nest."

Recipe courtesy of Klaus Kochendoerfer, Executive Chef, Grand Hotel Europe, St Petersburg

OPPOSITE: *Klaus Kochendoerfer, Executive Chef at St. Petersburg's Grand Hotel Europe, serves a number of vodka-infused dishes, including the vodka sorbet at left.*

NETHERLANDS

Vodkas from western Europe have also captured noticeable attention, particularly out of Holland. Ketel One originates from the Nolet Distillery, founded in 1691 in the Dutch port city of Schiedam. Ketel One claims to be one of the few vodkas being produced in pot stills.

ENGLAND

Although better known for its ales than for vodka, England is nevertheless the home of Black Death Vodka, an irreverent skull-and-crossbones brand which touts itself as "the vodka in bad taste."

UNITED STATES

Finally, we visit the United States, where the popular history of vodka goes all the way back to the 1950s. Until recently, the story of American vodka was dominated by such megacommodities as Smirnoff, Gilbey's, and Gordon's. But with the advent of ultrapremium import brands, American companies have responded aggressively. Superclean Skyy emerged from nowhere in 1992 to become one of America's best-selling vodkas. The United States has also seen other quality brands rise from the pack: Rain, Teton Glacier Potato Vodka, Rainbow Vodka, and winning the originality award, Tito's Handmade Texas Vodka.

CHAPTER 2

"It's funny, its colour is just like its smell. It's like that green you sometimes see in the heart of a white rose."

Somerset Maugham lauding the Polish vodka Zubrowka in his novel *The Razor's Edge*

SECRETS OF THE NEUTRAL SPIRIT

TRACING THE ORIGINS OF VODKA IS NO EASY TASK. A SUPPOSEDLY ODORLESS AND COLORLESS "NEUTRAL SPIRIT," VODKA HAS AN ELUSIVE PAST THAT EVEN THE MOST ASTUTE DETECTIVE WOULD FIND CHALLENGING TO DIG UP. ALL THE QUALITIES THAT MAKE VODKA A CLEAN, SMOOTH SPIRIT ARE ALSO THOSE CHARACTERISTICS THAT MASK ITS LINEAGE.

PAGE 44: *Vodka is served up at the GUM mall in Moscow.*
PAGE 45: *An industrial metal still from the Kaluga distillery in Kaluga, Russia.* **ABOVE:** *A frosty toast in Siberia with Baikalskaya Vodka.* **OPPOSITE:** *In making vodka, the icy "living" waters of Russia are never boiled or distilled, which is said to leave water "soulless."*

This vodka genealogy hardly mattered in the past. Until recently, vodka was . . . vodka. It was a commodity that eluded the brand-conscious. Vodka "call" drinks, where a bar patron specifies a certain brand of spirit, were unheard of until the 1980s, when Absolut and Stolichnaya took U.S. bars by storm. There emerged a need for brands to differentiate, and with this need came a heated debate on the origins of the spirit. It's a debate that's as cloudy as vodka is clear.

The other occupants of the bar shelf have distinguishing characters that make their histories easier to track down. With gin, juniper berries give us a clear fingerprint that leads to one Dr. Franciscus Sylvius, a Dutch physician who stumbled on the spirit while researching tropical diseases. The taste of

WHILE VODKA IS SUPPOSED TO BE, by definition, a neutral spirit, we all know there are tremendous differences among the premium brands. Every manufacturer has its own list of reasons why its vodka is the finest, making note of everything from the choice of raw materials to the filtration process. But recently distillers have begun touting the simplest ingredient, water, as the key to the product's character.

Given the fact that vodka producers are a notoriously nationalistic bunch, it's not surprising that many claim local water sources as the best for distillation. Perlova Vodka touts its extra-soft water from the Carpathian Mountains. Polish vodka maker Luksusowa claims to use water from artesian wells created during the Oligocene Epoch. Scope out the bottles in your favorite liquor store, and you'll see the extremes of the water boasts, from Kentucky wells to Dutch sand dunes.

Even the question of how water is prepared has seeped into this contest. For most vodkas, water is purified before it is used in the distillation process. Author William Pokhlebkin, in his definitive study *A History of Vodka,* praises the quality of Russian water and claims that "in no circumstances is the water subjected to boiling or distillation." Pokhlebkin proclaims this a point of national pride, stating, "This is an important distinguishing feature of Russian vodka, and one of the sources of its superiority." Distilling, he says, leaves a water "soulless" while Russian water is "living."

Of all the exotic water sources, none has matched that of Iceberg Vodka. This Canadian distiller has taken to melting Canadian icebergs for its new brand of vodka. And if aged water is what it takes to make the world's best vodka, this company may have the upper hand. The distiller's claim that the native icebergs, which yield 200 gallons (850l) of water for every ton of iceberg, are 12,000 years old.

rum smacks of warm tropical beaches, and it doesn't take long for our search to take us to Puerto Rico. Tequila comes from agave, and wine comes from grapes, telltale signs that lead to their origins. Some are even more obvious—it doesn't take a genius to figure out where scotch came from.

Then there's vodka, and most leads go cold. Only after a recent flurry of nationalistic vodka pride have clues begun to emerge. What better way to differentiate your country's vodka than to claim its heritage? From potatoes in Poland to rye in Russia, several countries want to claim vodka as their own.

ABOVE: *Even the folks in the lab deserve an elegant taste of the fruits of their labor (the Levize Distillery, St. Petersburg).*

OPPOSITE: *Although vodka is considered the most neutral of spirits, it is subjected to a surprising amount of chemical tinkering. Here, fruits and vegetables soak in jars of vodka at Infusion in San Francisco. The flavors (left to right): grapefruit, strawberry, vanilla bean, and lemon.*

SO WHAT IS VODKA, ANYWAY?

V odka is, for the most part, pure alcohol—at least in theory. It's manufactured from the fermentation of just about anything. The list of potential raw materials includes grain, corn, potatoes, sugar, and beets. But the question of which ingredient makes the best vodka is the source of much debate, and the fodder for endless marketing campaigns.

DISTILLATION

On paper, vodka making is a simple process. It starts with the liquid from a fermented "mash," which can come from almost anything organic. Many amateur distillers use plain sugar, bypassing the process that converts starches into sugar.

To create vodka, the mash is heated; this process takes advantage of a unique chemical phenomenon. Since alcohol evaporates at a lower temperature than water, distillers carefully monitor the temperature to boil off just the alcohol, which rises in a gaseous state. This gas travels through tubes or

columns and condenses into a low grade form of alcohol which is foul-smelling and chock-full of impurities. Since fermented mash was the basis for some of the most primitive alcoholic drinks, it's been speculated that distillation may have been discovered by someone accidentally heating this mash and having the bravado to drink the condensed vapors that came off it. Let it not be said that early distillers were without courage.

Other distillation methods involve a much higher temperature, boiling both the water and the alcohol together. The combined vapors steam into a large column. Because alcohol evaporates at a lower temperature than water, it follows that water condenses at a higher temperature than alcohol. This mixture of water and alcohol vapor cools as it rises into the column. In lower registers, water will condense, leaving the alcohol to condense farther up the column.

As an aside, one of the earliest methods of producing vodka utilized cold instead of heat to separate water and alcohol. Inhabitants of the northern regions of Europe would freeze fermenting liquid, which would solidify the water but leave the alcohol liquid. This method produced a greater-strength spirit than fermenting alone could, and was used until distillation became a common practice.

The product of a first distillation has been called "low wine," "simple wine," or even "moonshine." What makes this initial product so foul are impurities known as congeners, the most noticeable of which have some decidedly unpalatable names. The worst offenders are known as fusel fuels, which include butyl and isoamyl alcohol. While ancient spirits contained many of these contaminants, their elimination is the key to today's ultrapure vodkas. Ultimately, most distilleries strive to drive out everything but ethyl alcohol, the principal ingredient in vodka.

CHILLED GREEN GAZPACHO WITH CUCUMBER-INFUSED VODKA

INGREDIENTS LIST

4 cups diced tomatillos

2 tablespoons olive oil

2 yellow onions, diced

1 bunch cilantro, leaves only

6 garlic cloves, minced

2 jalapeño peppers, minced

½ cup cucumber-infused vodka*

¼ cup lime juice

1 avocado, diced

Crème fraîche or sour cream

Salt

Pepper

RECIPE

Sauté onions and tomatillos in 2 tablespoons olive oil. When soft, add garlic and jalapeños. Add 4 cups water, bring to a boil, and simmer for 15 minutes. Puree in a blender with the cilantro, lime juice, and vodka. Season to taste with salt and pepper. Chill. Garnish with diced avocado and crème fraîche.

*To make cucumber vodka: Put peeled, sliced cucumbers into vodka and let stand at room temperature for 10 days. Use about one part cucumber to three parts vodka.

Recipe courtesy of Infusion Bar & Restaurant, San Francisco, CA

In modern processes, the first distillate is further purified in a number of ways. First, it can be distilled yet again, to remove more of the congeners and produce a purer spirit. Some manufacturers distill their vodka up to six times. Some

VODA VS. WODA: The Vodka Name Wars

THE BATTLE OVER THE ANCESTRAL right to vodka has degenerated into name-calling, or at least name-claiming. Of the many contradictory claims that Russians and Poles have staked regarding the history of vodka, none is more at the center of the tornado than the conflict over the name itself.

Early alcoholic beverages were given lofty monikers that mimicked the heavenly feelings they invoked. Distilled spirits were called *aqua vitae*, translated as "the water of life." For many centuries, distilled spirits, including early vodkas, were simply called *vino* (from the latin *vinum*, meaning "wine"). As technology improved and the alcoholic content of these new spirits increased to the point where they could be lit on fire, they also became known in various parts of Europe and Asia as *bränntwein* or *goriachee vino* ("burning wine"), and as ardent water, from the Latin *ardere* ("to burn").

In *A History of Vodka*, author William Pokhlebkin explains that the word vodka first appeared in Russia in the fourteenth century but was not used regularly until the nineteenth century. The Russians claim that vodka is derived from the Russian word *voda*, meaning "water." Adding "-ka" to the end of a word creates a diminutive in Russia, much the way adding "-ette" reduces the scale of a noun in French. Thus, vodka in Russia is endearingly known as "little water," somewhat ironic given that country's huge appetite for the spirit.

In Poland, vodka is known as *wodka*, a derivation of *woda*, meaning "water." It is such a big part of that nation's history that a variation of this name, *wodki*, refers to every drink containing more than 20 percent alcohol, including rum and brandy. Vodka is also referred to in Poland as *czysta wodka* ("clear vodka"). Some claim that the existence in the early fifteenth century of the Polish words *okowita* and *okawita* ("water of life") predates Russian production by a century.

ABOVE AND OPPOSITE: *The origin of vodka is the fodder for many debates—and marketing campaigns. Both Russia and Poland claim that the very word "vodka" is derived from their respective words for water, and both back up their claim by producing many fine vodkas.*

While splitting hairs over wodka vs. vodka may seem petty, there's a lot at stake. In the era of superpremium vodkas, the ability to lay claim to the birth of vodka is important for business. As long as there are advertising budgets, there will be controversy. And while some say woda and some say voda, no one seems ready to call the whole thing off.

go so far as to take advantage of the different boiling points of different alcohols, using a specific distillation to remove, say, propyl alcohol and another to remove amyl alcohol. With each distillation, the alcoholic content increases.

But make no mistake, sometimes contaminants are darn tasty. While the goal of vodka production is to remove congeners, it should be noted that some so-called impurities are what give other liquors their character. Author Gordon Brown, in *Classic Spirits of the World,* states that many vodkas have about thirty milligrams of flavoring elements per liter while whiskies and cognacs have up to twenty-six hundred. The sting of tequila, the fruity nose of chardonnay, and the bitter taste of ale are all due to impurities.

This brings us to yet another idiosyncrasy of vodka. It is not aged. The minute the final product rolls out of a vodka distillery, it is ready to be consumed. Other liquors, particularly wine and scotch, are left for years in wooden casks, where they acquire the "qualities" of the wood. None of this goes on in vodka production. While other liquors get their identity from these outside influences, it's the lack of these influences that characterizes vodka.

POT STILLS

Early stills were decidedly low-tech compared to the massive stainless-steel columns in today's distilleries. The earliest forms of vodka were typically produced in single batches, from what is known as a "pot" still. The earliest pot stills were most likely made of earthen clay, with a simple apparatus off the top used to collect and condense the evaporating alcohol. Author Pokhlebkin doubts that the slow process of pot distillation occurred much before the fifteenth century, given the availability of grape wines and ales before that time.

Pot stills produced a less potent alcohol (typically 15–20 percent), and one that was hardly the clear, clean vodka we know today. Foul-smelling and murky, filled with poisonous fusel fuels, these liquids were often flavored with fruits, berries, and herbs (anything that would make them drinkable). This

practice of flavoring alcoholic beverages is ironic given today's trend toward flavored vodkas, popularized by Absolut's rainbow of vodka varieties (including Citron and Kurant). Another indication that we've come full circle in the flavor department is the current popularity of infused vodka. Infusion is the practice of soaking fresh fruits or vegetables in large dispensers of vodka

ABOVE: *"Proletariat of the world unite" reads this plaque outside a Moscow distillery.... "For achievements in the matter of Socialist construction we award this memorial flag to the collective of the Leningrad liquor-vodka plant—winner in the Socialist competition in honor of the 50th Anniversary of the Great October Revolution."* **OPPOSITE:** *While Russian vodka distillation continues to rely on traditional methods, so do their telecommunications. This antique phone, still used by workers, is located in the basement of the Levize distillery in St. Petersburg.*

for weeks at a time, resulting in a concoction of alcohol and vibrant flavorings.

The pot still was used for second or third distillations, each time yielding a stronger and purer alcohol. This turned out to be a slow and inefficient process, consuming huge amounts of fuel (some say it took almost 4 cubic yards [3m³] of wood to produce 25 gallons [100l] of vodka). In the early 1800s, the process known as "continuous distillation" became widely used thanks to an invention generally credited to Aeneas Coffey, an Irishman. The new device performed many distillations in a single process, allowing large batches of high-proof alcohol to be produced continuously and efficiently.

Continuous distillation is now the predominant process for producing vodka. Lately, we've been seeing a revival of the pot still method of distillation. One of the more visible proponents of this resurgence is Ketel One, which uses pot stills for the final distillation. This Dutch vodka producer claims to discard the first and last hundred gallons (425l) of each production, asserting that the ultrasmooth "heart" of the distillate can be captured only in pot stills.

FILTRATION

After the distillation process is completed, the resulting liquid is then diluted with water and put through a filtration process. This mechanical process is one of the most important distinctions between the different brands of vodka. It's also where many of the industry's trade secrets reside. Ask vodka distillers how they filter vodka, and the meekest of them will simply clam up. The bolder ones will tell you to get lost. Filtration is such a secret that, in many distilleries, the filtration rooms are strictly off-limits to visitors.

Early filtration processes were as simple as letting the spirit sit out while the solid particles settled to the bottom. As techniques improved, impurities were removed by using coagulants, such as milk or eggs, which solidified around the contaminants. As the vodka-making process was improved, many different materials were used for filtration, including cloth, wool, paper, sand, and crushed stone. Modern filtration procedures are much more active, involving many different and exotic materials. In the eighteenth century, charcoal was found to be an effective filtering agent. Today, the most common ingredient used in filtration is activated charcoal, which is particularly good at removing oily contaminants.

Even though filtration remains a trade secret, inevitably some details from the filtration room do make it to the marketing department. Today you see many vodka makers talking about their filtration processes. Some of the more interesting are outlined in the table below.

MANUFACTURER	COUNTRY OF ORIGIN	FILTRATION MEDIUM
EVERCLEAR	UNITED STATES	"Select American charcoal"
FINLANDIA	FINLAND	Glacial rock
KETEL ONE	HOLLAND	Sand from nearby dunes
LUKSUSOWA	POLAND	Charcoal and oak chips
PERLOVA	UKRAINE	Quartz crystal and birch charcoal
RAIN	UNITED STATES	Limestone
STOLICHNAYA	RUSSIA	Fine quartz sand and birch charcoal
TANQUERAY	ENGLAND	Red ochil granite chips
TETON	UNITED STATES	Charcoal and garnet crystals

One manufacturer, however, says "nix" to the whole filtration process. In *The Martini Companion*, authors Gary Regan and Mardee Haidin Regan quote Lars Nellmer, marketing director for the Absolut Company, as saying, "Charcoal is for barbecues. We rely on our distinctive distillation methods to produce a pure product that still bears character."

Once the filtration process is complete, water is often added to dilute the spirit to the acceptable strength. Vodka is typically 80 proof (40 percent alcohol). While some brands are

ABOVE: *Bottles of ultrapure vodka roll off the assembly line in a Siberian distillery. Technically, other spirits, such as Scotch, get their color and taste from impurities; vodka earns its shimmering character from the lack of them.*

as much as 100 proof, it's rare to find a vodka more potent than that. The European Economic Community stipulates that a vodka be at least 75 proof.

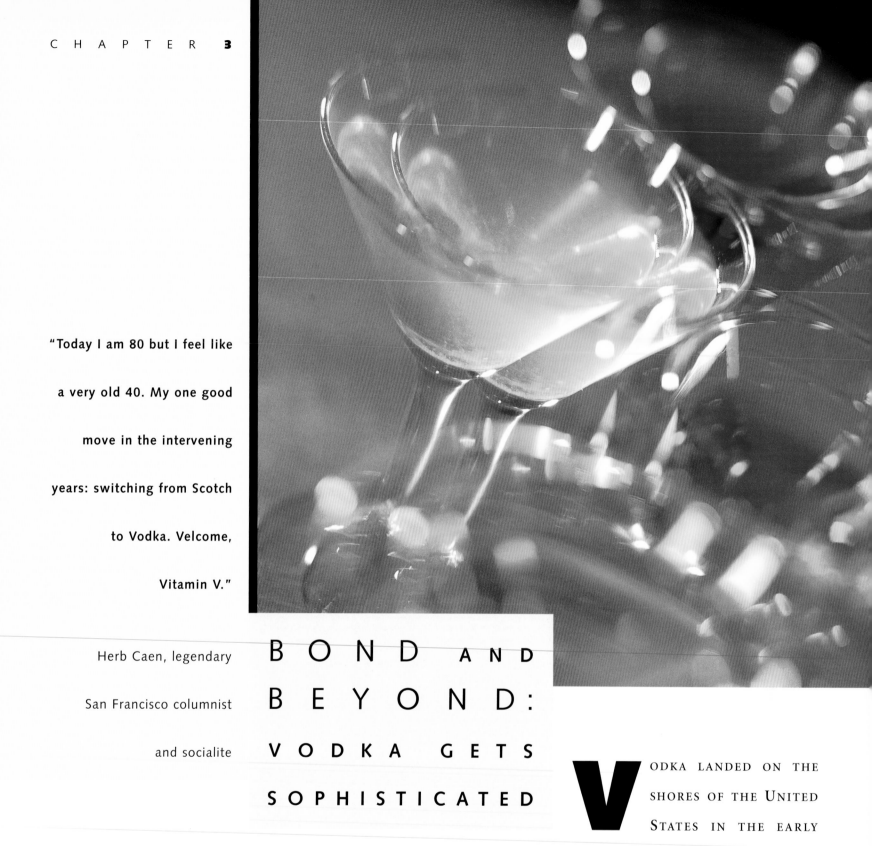

CHAPTER 3

"Today I am 80 but I feel like a very old 40. My one good move in the intervening years: switching from Scotch to Vodka. Velcome, Vitamin V."

Herb Caen, legendary San Francisco columnist and socialite

BOND AND BEYOND:
VODKA GETS SOPHISTICATED

VODKA LANDED ON THE SHORES OF THE UNITED STATES IN THE EARLY PART OF THE TWENTIETH CENTURY, AND A QUIET LANDING IT WAS. WHILE EUROPEAN CULTURE WAS DEEPLY SOAKED IN VODKA, AMERICAN HABITS WERE DECIDEDLY AMBER-TONED. AMERICA HAS LONG BEEN A LAND OF WHISKEY DRINKERS, AND THE "BROWN SPIRITS" HAVE A LEGACY IN THE STATES THAT GOES BACK TO THE REVOLUTIONARY WAR.

A few American presidents had a thing for whiskey. George Washington is said to have cultivated rye at Mount Vernon. Abraham Lincoln, though not a man to tip one back, appreciated the assumed value of whiskey during the trying times of the Civil War. When someone bemoaned the whiskey-loving ways of General Ulysses S. Grant, Lincoln heralded the general's battlefield successes and suggested that Grant's favorite brand of whiskey be doled out to the other Union generals.

Americans had an exclusive love affair with whiskey until the 1920s. The smooth, aged taste of Kentucky bourbon, in particular, emerged as a national favorite. Unfortunately, the establishment of Prohibition in 1919 brought an abrupt end to this romance. The powers that be speculated that "The Noble Experiment" would wean the country off alcohol, but it did little more than shift attention to a different kind of hooch.

With the threat of G-men pounding down the door, Americans turned to more expedient forms of alcohol production. "Clear" spirits, which were easier to produce and did not have to be aged, started popping up. A society of whiskey drinkers added the phrase "bathtub gin" to their vocabulary to describe their homemade spirits, and the distinctive juniper berry smack of gin soon found its way onto American palates. Other spirits also took this as their entrance cue, and rum from the Caribbean arrived on the scene. By the time Franklin D. Roosevelt said "uncle" on prohibition in 1933, the United States had acquired a distinct taste for neutral spirits, gin in particular.

CLASSIC DRY VODKA MARTINI

INGREDIENTS LIST

8 parts vodka

1 part extra-dry vermouth

RECIPE

Pour over ice, shake or stir, strain, and serve in a martini glass with an olive.

RUSSIAN VODKA HITS THE ROAD

Vodka, however, took a more circuitous route to American taste buds. Its journey began with royal roots in Russia, where in 1886 Piotr Smirnov was awarded the warrant to supply vodka to the imperial court of Tsar Alexander III. The Russian revolution in 1917 brought an end to the family business and Vladimir Smirnov, a family member, took possession of the brand and recipe and left the country, narrowly escaping the firing squad in the process. But establishing the brand outside of Russia proved to be an enormous challenge. Vladimir landed in Istanbul, then Poland, and finally France. In 1928 he opened a small distillery but the reception to his vodka, by then known as Smirnoff, was less than overwhelming.

Unsuccessful in his attempts to resurrect the brand, Vladimir sold the formula and the name to Rudolph Kunett, whose family had supplied grain to the Smirnovs before the revolution. Kunett's venture failed as well. By 1939 Kunett had managed to sell only six thousand cases a year, and he, too, sold

SMIRNOFF IN THE FIFTIES:
The Remedy for the Three Martini Lunch

WHILE TODAY'S ULTRAPREMIUM vodka drinker prefers quality over quantity, this wasn't always the case. In 1950s America, when the words "politically" and "correct" were rarely uttered in the same sentence, business executives were known to toss back a few in a ritual that became known as the "three martini lunch." The swinging cocktail culture of the time tolerated (and sometimes encouraged) the habit, but protocol still required many executives to keep their noontime nips to themselves when they returned to the office.

Enter Smirnoff Vodka. Originally marketed as the "white whiskey," Smirnoff was having a hell of a time getting noticed in America. The company stumbled badly with an original marketing campaign that stated "No taste. No smell." Sales were low, and a new slogan was needed.

Smirnoff debuted a new marketing campaign around the slogan "It Leaves You Breathless," and promoted a Smirnoff cocktail as the one to "start with and stay with." Ads featured such stars as Benny Goodman and Harpo Marx, with copy claiming Smirnoff "leaves no whisper of liquor on your lips."

Whether deliberate or not, Smirnoff had stumbled upon a message that was music to the ears of the lunchtime cocktail crowd. Vodka's neutral character was just the kind of stealth the partaking executive craved. Soon, vodka martinis started outselling gin, and that helped catapult Smirnoff into a powerhouse brand. By 1976, vodka had overtaken whiskey as the nation's most popular spirit, with Smirnoff leading the way.

Today's cocktail lovers embrace the martini as much as their counterparts of the past, but in a very different way. Competitive workplaces have made the three martini lunch a relic of the past, and most connoisseurs are now looking for swank brands, not large portions. To its credit, Smirnoff has swung with the times. The venerable brand now leaves their customers breathless with Smirnoff Black Traditional Russian Vodka, a triple-distilled vodka straight from the motherland.

ABOVE: *Vodka at the bar at Hotel Baltschug Kempinski in Moscow.* **OPPOSITE:** *Vodka has historically been the toast of Moscow, but it was originally a tough sell in the United States. However, when a copper mug and an overstock of ginger beer combined with vodka to make the "Moscow Mule," the neutral spirit caught on in the States.*

out. A then-tiny liquor company, G.F. Hueblein & Brothers bought Smirnoff for a mere $14,000 (plus royalties), determined to introduce vodka to the United States.

WHITE WHISKEY

It was an uphill struggle. Marketed as "white whiskey," vodka was portrayed in early advertisements as having "No taste. No smell." In the United States that translated into no sales, and vodka languished on the retail shelf for years. It was not until a fateful day in 1946 when, as the result of a rare combination of an emerging cocktail culture and an overstock of ginger beer, vodka finally received a proper introduction to American culture.

ABOVE: *Truly an international brand, Znaps is distilled in England from Swedish raw materials. As vodka borders fall, cross cultural trends are becoming more prevalent.*

OPPOSITE: *Active cultures play a role in fermentation, and distillery workers keep an eye on their own photosynthesis lab to make sure that the cultures are correct.*

Legend has it that Hueblein executive John Martin, struggling to sell vodka in America, found a soul mate in Jack Morgan, owner of a Los Angeles bar called The Cock and Bull. It seemed that Morgan also had a dud on his hands, an obscure product called ginger beer. The two got together and concocted a drink of ginger beer, vodka, and lime, called it the Moscow Mule, and served it in a copper cup. Mixed drinks

FETTUCCINI WITH DUNGENESS CRAB WITH SKYY VODKA ESSENCE

INGREDIENTS LIST

1 pound fettuccini pasta

8 ounces cooked Dungeness crab meat

¼ cup sliced shallots

2 sprigs thyme (fresh)

1 bay leaf

6 ounces butter

1 teaspoon olive oil

2 tablespoons chopped chives

8 ounces Skyy Vodka

Salt and pepper

RECIPE

In a stainless steel pot place olive oil, shallots, bay leaf, and thyme and sweat over low heat for about 7 minutes. Make sure there is no brown color in the pot. Add the Skyy Vodka and reduce heat by two thirds.

Whisk in the butter slowly until emulsified.

Strain the sauce through a fine strainer, then add the chopped chives to the sauce.

Cook the pasta in salted boiling water until al dente.

Drain the pasta, place in a mixing bowl, and add the sauce and the cooked, picked crab meat. Ready to serve.

Serves 4

Recipe by Rabah Abusbaltan, Chef de Cuisine, One Market Restaurant, San Francisco, and courtesy of Skyy Spirits

OPPOSITE: *Originally used in the 1950s as the alcoholic kick for decidedly lowbrow drinks, vodka has been elevated to the elixir of choice for top-shelf cocktails, like those served at Voda, the swank vodka bar in Los Angeles.*

were taking off in the States, and Martin and Morgan were convinced they had a hit.

While Morgan promoted the Moscow Mule to his patrons at The Cock and Bull, Martin hit the road with a slick new device that had just hit the market, the Polaroid camera. With camera, vodka, and copper mugs in hand, Martin toured the bars of New York City, where bartenders were thrilled to be photographed with this new device. Of course, they were always seen with a Moscow Mule in their hand. The drink became a hit and opened the door for vodka in the United States. It also ushered in a new era in advertising. "For the first time, an invented cocktail was being used as a marketing device," claims William Grimes in his book *Straight Up or On the Rocks: A Cultural History of American Drink.*

THE COCKTAIL ERA

The 1950s brought swank and sophistication to America, and with them the love of cocktails. It also brought Joseph McCarthy and conformity, and vodka became an ideal spirit for the time. Alcoholic drinks became the social icebreaker that no one dared pass up, and vodka's anonymous

"Happiness is a tub covered by a pot."

Ancient proverb praising the virtues of

the home still

INFERNO PENNE ROMANOFF

INGREDIENTS LIST

¼ cup extra-virgin olive oil

4 large, fresh garlic cloves

One 28 ounce can stewed Italian plum tomatoes

1 pound dried penne (fresh is better, but harder to find)

3 tablespoons Inferno Pepper Pot Vodka

1 cup heavy cream

¼ cup flat-leaf parsley leaves (snipped with scissors)

Salt to taste

RECIPE

In an unheated skillet large enough to hold the pasta later on, combine oil, garlic, and a pinch of salt, stirring to coat with oil.

Cook over medium heat until garlic turns golden. Add crushed stewed tomatoes. Stir and simmer uncovered, until sauce thickens. Taste for seasoning, and adjust as required.

Boil pasta until al dente.

Add drained pasta to skillet. Toss. Add Inferno Vodka, toss again, then add cream and toss once more. Cover, reduce to low heat, and let rest for a few minutes to allow pasta to absorb sauce.

Add the parsley, then toss again. Serve.

Note: Traditionally, cheese is never served with this dish. If you prefer cheese, however, parmigiano reggiano is recommended.

Serves 6–8

INFERNO GUACAMOLE

INGREDIENTS LIST

2 ripe avocados

1 tablespoon finely chopped onion

1 tomato, peeled, seeded, and chopped

1 teaspoon finely chopped coriander

½ teaspoon salt

2 tablespoons Inferno Pepper Pot Vodka

RECIPE

Peel, chop, and mash avocados to a puree, then add all other ingredients and mix well. Cover with aluminum foil to prevent discoloration, and refrigerate. Serve with raw vegetables or toasted tortillas.

Makes about 2 cups.

Recipes courtesy of Kittling Ridge Estate Wines & Spirits, makers of Inferno Pepper Pot Vodka

OPPOSITE: *For those who dare, there is Inferno Vodka, made with the distiller's own special, hydroponically grown, red-hot "911" peppers. Good for sipping, good for cooking with.*

JAMES BOND: The Art of Cool and the Vodka Martini

DEBONAIR AND DANGEROUS with the women, lethal to the Communists. In the 1950s and 1960s, these were the benchmarks for the Western man. While the majority of men spent the postwar years in more domestic pursuits, this only intensified the image of the alter ego: the gun-toting globe-trotter who foiled the Reds and bedded the dames (in between cocktails, of course).

In 1953, Ian Fleming's novel *Casino Royale* brought the brash and the sophisticated together in secret agent James Bond. Bond was the perfect combination of danger and flair that the times required, and vodka was right by his side. The secret agent's missions often took a back seat to cocktails, his first concoction being vodka, gin, and Kina Lillet (a French aperitif). Bond liked his drinks "large and very strong and very cold and very well made." He named the drink "The Vesper" after the lovely double agent he encountered.

Why gin and vodka together? Some say it was a natural, given how the East and West mixed it up during the Cold War. But some people have more romantic notions regarding the combination. In his classic book *The Martini,* author Barnaby Conrad III asks, "Did Fleming mean the gin and vodka to have symbolic rapport with the double agent?" Possibly. After Vesper bit the dust, the drink was buried with her.

In 1962, when agent 007 hit the big screen in *Dr. No,* vodka was the spirit of choice, and bottles of Smirnoff seem to appear in every scene. Bond swizzles vodka martinis at every turn, and even his adversaries oblige him. After being captured and drugged by the notorious Dr. No, Bond is offered a cocktail by the Chinese madman. "Medium dry martini, lemon peel," the good doctor proudly proclaims. With a nuclear reactor below him and an arsenal of machine guns pointed at his gut, Bond shows his priorities. "Vodka?" he asks his adversary. "Of course," the doctor replies.

Later, Bond would branch out in the alcohol department, sipping bourbon, champagne, and even Americanos. But it's his early vodka endeavors, and his insistence that a martini be "shaken, not stirred," that serve as his cocktail legacy.

nature served as the perfect ingredient for what became known as "mixed drinks." In *The Martini Companion,* authors Gary Regan and Mardee Haidin Regan make a good point about this post-war cocktail peer pressure. "Cocktails were de rigueur," they remind us. "Vodka appealed to the many people who wanted to be part of the scene but weren't enamored of the taste of liquor."

With its conveniently neutral nature, vodka became the main ingredient in a long list of popular mixed drinks. In 1951, one cocktail book had the audacity to break tradition and suggest a vodka martini. The "Vodkatini," as it was described in Ted Saucier's cocktail book *Bottom's Up,* called for vodka,

OPPOSITE: *The Vodka martini, or "Vodkatini," as it was first known, started humbly as the obscure cousin of the classic gin martini. Today, the classic vodka martini outsells its gin counterpart in the more upscale clubs by as much as five to one.*

vermouth, and lemon peel. Today, vodka martinis outsell gin martinis by almost five to one.

Manufacturers of vodka started to encourage experimentation and featured mixed drinks in their advertisements. Smirnoff encouraged the Moscow Mule. The Bloody Mary had

been around since the 1930s but it became a national hit after George Jessel was seen swilling one in a 1955 advertising campaign. This tangy combination of vodka, tomato juice, Worcestershire sauce, Tabasco, lemon, and celery salt became known as a hangover remedy for the morning after.

The Moscow Mule and the Bloody Mary opened the floodgates for a long line of vodka cocktails, and soon the neutral spirit became the base for a whole legion of household drinks. And what was more universal in households than orange juice? A mating was arranged, and the infamous Screwdriver was born. Soon things got a little silly, though, and vodka began a slow, steady decline into parody. The White and Black Russians married vodka with Kahlua, forming a popular dessert-like drink, the antithesis of cocktail chic.

Still, it took a California surfer named Harvey to bring vodka to new lows. Harvey was taken to supplementing his Screwdrivers with Galliano, a concoction that supposedly left him bouncing off barroom walls. The surfer's sweet concoction was dubbed a Harvey Wallbanger, which eventually hit the cellar of respectability, taking vodka along with it. "Vodka was definitely not a society drink," says marketer Michel

PAGE 78: *Science or art? Great vodka is a mix of tradition and technology at the Znaps distillery in Birmingham, England, where vodka is made with Swedish ingredients and methods.*
PAGE 79: *Inferno Vodka's main still in Ontario, Canada.* **BELOW:** *Bottle caps awaiting bottles, and a few that have finally met their counterparts.*

Roux, the trailblazing CEO of Carillon Importers who is credited with Absolut's invasion of America and now imports Stolichnaya. "It was peddled by the pint and the half pint in the wrong places."

During this Cold War cocktail era, the premium vodkas of Russia and Poland, so familiar to us today, were unheard of. In his 1961 book *The Fine Art of Mixing Drinks,* author David Embury could not even comment on eastern European vodka for lack of samples. "There is no Russian vodka available in the United States at the present time," he writes, lamenting, "I have had no Russian or Polish vodka for quite a number of years." Poor guy.

ABOVE: *Two of Russia's greatest delicacies, vodka and caviar, are combined to create the unique sauce that tops Heinrich Moderle's pan-fried sturgeon at the Hotel Baltschug Kempinski, in Moscow.*

CHAPTER 4

"The vodka is good, but the meat is rotten."

An erroneous English-to-Russian-to-English computer translation of "The spirit is strong, but the flesh is weak." From Raymand Smullyan's *What Is the Name of This Book*

ЛИВИЗ

100 лет

40%

0,7л

BACK to the TOP:
THE VODKA REVIVAL

THE 1970S WERE THE RODNEY DANGERFIELD YEARS FOR VODKA IN THE UNITED STATES. TUTTI-FRUTTI CONCOCTIONS, LED BY THE HARVEY WALLBANGER, BROUGHT LITTLE RESPECT TO A SPIRIT THAT BECAME FIRMLY ENTRENCHED IN A SUPPORTING ROLE. MANY HISTORIANS LAMENT THIS PERIOD AS ONE IN WHICH SOPHISTICATION AND STYLE TEETERED ON THE EDGE OF OBSCURITY.

For many, the seventies represented the nadir of American social grace. Disco and drugs, combined with Watergate and Vietnam, yielded a lethal cocktail that emphasized street over sophistication. Aged spirits and meticulous distillation meant little in this age of excess.

VODKA IN THE GO-GO EIGHTIES: A MARKETING REVOLUTION

As the eighties approached, the United States began to slowly awaken from its cultural shock and some signs of refinement could be observed. While the history book is still open with regard to his political legacy, few can doubt Ronald Reagan's black-tie influence on the nation. He and his wife Nancy were as conservative in their style as they were in their politics. When they relocated to the White House in 1980, many say that 1600 Pennsylvania Avenue experienced a type of pomp and circumstance not seen since the Kennedy years.

Stolichnaya was the first Russian vodka immigrant in the 1950s. But it was not until the 1980s that "Stoli" got the marketing muscle it needed. Ads proclaimed a superior vodka made from winter wheat and glacial water, and people started listening. Sales took off, and in 1989, the company introduced a superpremium brand extension. The first shipment sold out almost immediately.

But it took an obscure, Swedish company to bring premium vodka into superstardom. The V&S Vin & Sprit AB was preparing for the centennial anniversary of Absolut Rent Brännvin ("absolutely pure vodka"), a product first produced by the Swedish "king of vodka," Lars Olsson Smith, in 1879. It made the decision to export a new Absolutely Pure Vodka, one that played up a superclean image. The name: Absolut.

Strong, bold and modern.

 # (Printed directly onto) Blue and copy - for bottle.

Bottle has a matt finish.

PAGE 82: *The Levize distillery's own house vodka.*
PAGE 83: *A classic vodka martini, made just the way James Bond would want it.* **ABOVE:** *A sneak peek at the next generation: a vodka marketing plan for an as-of-yet unnamed new Swedish brand.* **OPPOSITE:** *The torch of fine tradition—with a colorful, modern twist—crosses the table every day at Pravda in New York City.*

ABOVE: *An antique vodka bottle from the nineteenth century. The long tradition of discerning vodka packaging has enjoyed a revival of late, with Absolut's distinctive medicine bottle leading the way.* OPPOSITE: *Trying to keep up with the demand at the Inferno Vodka's bottling facility. By the early eighties, vodka had successfully made the transition from commodity spirit to connoisseur's choice.*

A brilliant marketing campaign, centered around the tag line Absolut Perfection, captured the essence of the brand. Other ads in the same vein, such as Absolut Heaven, followed. The product packaging was as enticing as the vodka within. The company adopted the design from an old Swedish medicine bottle found in an antique shop (fitting, since most of the early uses for distilled spirits were medicinal) and complemented it with a medal around the neck, which featured the founder. In 1979 Absolut won a major award for best product package and by 1985 it was leading imported vodka in the U.S.

Michel Roux, who handled Absolut at the time, is the man credited by many for blazing the trail for premium vodka. Roux saw a shift in the market and seized on it. "The scotch business started to decline, and customers started to move to white goods," he says today. "There was an upscale opportunity for vodka."

The next step would take Absolut from a popular brand to icon status. In 1985 renowned artist Andy Warhol was commissioned to do a painting of the Absolut bottle, and the resulting work was featured in advertising throughout the world.

Like he'd done with a Campbell Soup can, Warhol had transformed Absolut into pop art, a cultural phenomenon that fascinates buyers to this day.

Why combine art and vodka? Michel Roux says it was a natural. "Alcoholic beverage advertising is pretty boring," he says pointedly. "Something had to be done." A series of other Absolut commissions followed, featuring art by the likes of Edward Ruscha and Keith Haring. Roux knew the association would pay off, as it continues to do today. Now spearheading the marketing for Stolichnaya, Roux continues to build bridges with the art world through Stoli's "Freedom of Vodka" campaigns featuring Russian artists. "Using art to do advertising gives the aura of prestige and culture," Roux claims. "It takes the stigma away."

ABOVE: *A cozy corner at The Caviar Bar in the Grand Hotel Europe, in St. Petersburg, built in 1824.* **OPPOSITE:** *Always a main ingredient behind the bar, vodka soon caught the attention of discerning chefs, and now has a permanent place for itself in the finest kitchens. Here is Heinrich Moderle's take on the other Russian delicacy, borscht, served up in style at Hotel Baltschug Kempinski.*

SKYY TROPICAL SUNSET

INGREDIENTS

2 tablespoons lemon juice

2 tablespoons orange juice

2 tablespoons blood orange juice

2 tablespoons mango juice

2 tablespoons key lime juice

2 tablespoons pink guava juice

2 tablespoons raspberry juice

2 tablespoons coconut juice

½ cup sugar

8 tablespoons Skyy Vodka

2 cups heavy cream

RECIPE

Whip the heavy cream to medium peaks. Gradually sprinkle in the sugar. Next, slowly pour in the Skyy Vodka.

Divide the whipped cream into 8 portions. Add one type of fruit juice to each of the separate creams.

Layer each mousse into a martini glass. Chill.

Garnish with candied lemon zest and serve.

Serves 8

Recipe courtesy of Skyy Spirits

Absolut followed this campaign brilliantly, combining the word Absolut with just about anything: Absolut Fashion featured famous models of the day; Absolut Holiday brought vodka to the Yuletide season's celebrations. As a result, premium vodka was left permanently etched into the American psyche.

But advertising might not be the whole story behind Absolut's success. The early 1980s were a time of renewed

OPPOSITE: *Vodka and caviar sit packed in ice, waiting for the evening's festivities to commence at the Europe Restaurant, St. Petersburg's most elegant dining choice. The stained glass real wall was executed in 1905.*

Cold War sentiments. Reagan's military buildup, the war in Afghanistan, and the downing of Korean Flight 007 all combined to create a decidedly anti-Russian mood in the United States. Since vodka was so closely associated with Russia, Swedish Absolut provided the perfect way to say *nyet* to the Reds.

BACK TO THE LOUNGE: THE 1990S VODKA RENAISSANCE

In the early 1990s, a cocktail/lounge revival swept across the United States. Several factors contributed to the return of swank, from baby boomers hitting their peak earning years to Generation X-ers in their twenties becoming tired of the angst and alienation of "grunge" style. Premium cigar sales hit an all-time high, and single-malt scotch became all the rage. Classic barware and vintage swing recordings, once selling for pennies at flea markets, became rare collectibles with lofty price tags.

One of the most noticeable signifiers of this new era was the return of the venerable martini. Once your old man's cocktail, the martini took on new status with a younger generation, and pulled vodka along with it. Although the gin martini is the stuff of purists, more vodka martinis now move across American bars. The result? "Vodka got more prestige and sophistication," says Michel Roux. Today, the martini has regained its mythical magic. "For me, the Dry Martini remains an American symbol of elusive perfection, a kind of pagan

Holy Grail," wrote Barnaby Conrad III in *The Martini*, his 1995 book, which is widely credited as a major influence in the martini revival.

Still not convinced of vodka's current migration to the top shelf? Let's look at the numbers. According to the United States' Distilled Spirits Council, the total amount of vodka (domestic and imported) entering U.S. trade channels decreased in the past decade, from over 95 million gallons (400 million l) in 1986 to just under 92 million in 1995 (the latest year that figures were available). However, imported vodka's share of that number doubled in the same period, from 6 percent to over 12 percent. Since 1980, the amount of vodka imported into the United States has skyrocketed, from 1.46 million gallons (6.2 million l) in 1980 to more than

10 million gallons (42 million l) in 1996. Vodka imports are now double that of gin. While not all imported vodka is top-notch stuff, the trend shows a high fascination with more exotic and sophisticated spirits.

The number of premium vodkas being marketed in the United States has also increased. Noteworthy newcomers, such as Ketel One from Holland and Skyy in the United

ABOVE: *Plant manager/inspector at the Znaps distillery. Znaps is billed as a Swedish style vodka, made from Swedish ingredients. But it is distilled in Birmingham, England.* **OPPOSITE:** *A sparkling taste of Russian vodka in a cut-crystal glass.*

VODKA-GRENADINE JELLY WITH EXOTIC FRUITS

INGREDIENTS

1 cup vodka

1 teaspoon lemon juice

5 gelatin leaves

1 cup grenadine syrup

Lychees

Lime

Figs

Mango

Physalis

Grapes

Kiwi

Chocolate flakes

RECIPE

Soak all the gelatin leaves in cold water to soften. To make vodka jelly, dissolve half the softened gelatin leaves in warm lemon juice, then add vodka and mix. To make grenadine jelly, in a separate bowl, dissolve the remaining gelatin leaves in warm grenadine.

Pour a thin layer of the vodka mix into a mold, and refrigerate until it solidifies. Pour a layer of the grenadine mix on top of the vodka gelatin, and refrigerate again until solidified. Continue pouring and chilling, alternating layers of the vodka and grenadine gelatins until the mold is filled, allowing the shape and size of the mold to dictate the total number of layers.

Refrigerate completed mold for 3 or 4 hours. When properly chilled, arrange molded jelly form on a plate with the exotic fruits and the chocolate.

Recipe courtesy of Heinrich Moderle, Executive Chef, Hotel Baltschug Kempinski, Moscow

OPPOSITE: *Heinrich Moderle, chef at the Hotel Baltschug Kempinski in Moscow, just across from the walls of the Kremlin.*

AMERICA'S VODKA MAKER: SKYY'S MAURICE KANBAR

How One Man's Headache Became One Great Vodka

MAURICE KANBAR HAD A glass of cognac, and got a headache. No big deal, right? Most of us would take an aspirin and forget about it. In fact, that's just what Maurice's physician friend told him to do. But that was not good enough for Maurice. He wanted to know why he had gotten a headache. His friend rolled his eyes and told Maurice that most alcohol contains impurities known as congeners, which some people are sensitive to. That got Maurice thinking about a vodka that would be totally void of congeners, one that was as pure as possible. "They thought I was nuts," he says today.

A native New Yorker whose rapid-fire speech demonstrates his intensity and drive, Maurice Kanbar has more than thirty patents to his credit, including many medical devices and a sweater comb known as the De-Fuzz It. In 1971 Kanbar opened the nation's first multiplex in New York City to satisfy his passion for films. This is definitely a guy who solves his own problems.

"If someone would make a dependably pure vodka," Kanbar recalls saying at the time, "it's the only vodka I would drink." But realizing no one else was that fanatical, Kanbar spent the next few years researching how to make a superpure vodka, void of any congeners. By 1992, Kanbar had the formula and bottled his concoction in a striking cobalt-blue bottle. Then he gave it an equally simple name: Skyy Vodka. But would it sell?

"All my friends told me I was completely crazy to go up against the big monsters of the alcohol business," he says. As it turned out, he wasn't. Distilled four times and produced with a proprietary three-step filtration process, Skyy flew off the shelves. Today, Kanbar claims it's the third best-selling premium vodka in the United States, behind Absolut and Stolichnaya.

Even after all this success, the memory of the headache must still be with Kanbar. He's obsessed with purity. Sitting in his Victorian office in San Francisco, Kanbar excitedly shows off a laboratory analysis of his "no hangover vodka." His report displays a list of impurities with nasty-sounding names like "amyl alcohol." Next to each impurity it reads "less than 1 milligram per liter." Kanbar leans back and smiles, knowing he's satisfied his toughest customer: himself.

Today, many products are the result of impersonal market research, and Kanbar scoffs at such hands-off tactics. While other companies wonder if *they'll* buy a product, Kanbar asks, "Who the hell are *they?*" His eleventh commandment: "Thou shalt not bullshit thyself."

PAGE 96: *The ornate bar at the Four Seasons Hotel in Los Angeles specializes in "classic cocktails which time and trends made us forget."* **PAGE 97:** *Four Seasons Executive Chef Carrie Nahabedian oversees the kitchen.* **OPPOSITE:** *With a fanaticism for purity, Skyy Vodka, based in San Francisco, came out of nowhere to become one of the best-selling vodkas in the United States, and the top-selling premium domestic brand.*

With today's vodka makers working overtime to improve the quality of their spirits, and bar-stool consumers demanding superpremium brands, it's safe to say that we've come a long way from the hooch-swilling, Harvey Wallbanger days. But if there's any doubt remaining that we've entered a whole new vodka era, consider Rainbow Spirits. Named after the universal symbol for "hope" in the gay and lesbian communities, the company donates half of the net profits from its Rainbow Vodka to nonprofit AIDS organizations. "The liquor industry is considered the bad boy," says Michael Avitable, Rainbow's director of marketing. "Rainbow was started to give back to the community." In addition to its triple-distilled grain vodka, Rainbow also produces rum, gin, tequila, and triple-sec.

From crude, medieval stills to royal palaces, through revolutions and prohibitions, vodka has come a long way. But where does it go from here? Is there no stopping vodka's ascent to the height of culture? While vodka marketers search for the next brand extension, some have even higher aspirations. "I wish everyone did with vodka what Andy Warhol did when he was alive," Michel Roux jokes. "He used it as perfume."

ABOVE: *Deep within vodka country near Moscow lies Lake Baikal, the largest, deepest lake in the world.*
OPPOSITE: *A distillery worker in Kaluga, Russia.*

States, are giving Stolichnaya and Absolut a run for their money, and established mainstream brands such as Smirnoff, are now marketing top-shelf varieties. Today's wise vodka companies see the writing on the wall, and aren't going to be left out in the cold. If a new generation is putting on the dog, sporting a premium cigar, and heading to the clubs for a fine cocktail, vodka makers want to have plenty of premium spirits there to greet them.

RUSSIA TODAY: THE COCKTAIL REVOLUTION THAT NEVER OCCURRED

Today, the United States enjoys a period of unparalleled prosperity and personal responsibility, as evidenced by the growing popularity of premium vodkas. Unfortunately, this is not the case throughout the world. While Americans marimba their way through a cocktail period crooning "quality not quantity," the situation is much different in Russia. As close as vodka is to Russian society, it appears closer still to its country's societal ills.

Vodka has long been a refuge from the drab reality of Russian life, its sedating effect providing the antidote for decades of authoritative rule and bureaucracy. Russian consumption of vodka increased dramatically after World War II, during which the Red Army had provided vodka rations to military troops. Ultimately these rations had a devastating effect on the military and, eventually, the public. Historian William Pokhlebkin notes, in *A History of Vodka,* "A profound change had occurred in the psychology of people who until this time had regarded drunkenness as something shameful."

The price of vodka was kept low through the 1950s and 1960s, creating a culture of consumption. By the 1970s, drunkenness on the job had become, according to Pokhlebkin, an "everyday phenomenon." Soviet leader Konstantin Chernenko, said to have a voracious appetite for vodka, died from what is rumored to have been an alcohol-related illness.

PAGES 102 AND 103: *Russian distillery workers, on the job, and on their break. In many Russian distilleries, men—with their voracious appetites for vodka—are prohibited from working anywhere near the stills, and can work only around sealed cartons or, as seen here, empty bottles. The higher-paying jobs along the production lines are available to women only.*

OPPOSITE: *Distillery tanks in Siberia, Russia. Vodka has historically been a double-edged sword in Russia: an important source of state income while contributing to many of the society's problems.*

Finances played no small part in the Soviet Union's vodka plans. Under Communist rule, proceeds from alcohol duties and sales, known as *pyanye dengi* ("drunken money"), accounted for almost one third of all state revenue; it's been argued that vodka actually financed most of the USSR's budget for much of the twentieth century. Attempts at control have even rocked commodity markets: the production of samogon, a fiery home-brewed concoction of sugar, water, and bread or potatoes, has accounted for 10 percent of the nation's sugar consumption in times of tight vodka controls.

The 1980s saw one last attempt at curbing the nation's thirst for vodka. Under Mikhail Gorbachev, an all-out campaign against alcoholism was waged. Propaganda against the "demon vodka" warned against the dangers of drunkenness, and for a while it actually worked. By 1986, consumption had decreased 26 percent, but state revenues from alcohol plummeted. So famous was Gorbachev's crackdown on vodka, a then-popular joke was told that pointed up inherent conflicts and went something like this: The Soviet leader exploded in anger when he found his glamorous wife Raisa in bed with one of his ministers. The surprised minister bolted out of bed and exclaimed, "Why are you so upset? We weren't drinking vodka."

WARM OYSTERS WITH SKYY VODKA LEMON ESSENCE

INGREDIENTS

24 Long Island oysters

½ cup Skyy Vodka

1 tablespoon Meyer lemon juice

½ tablespoon cracked black pepper

2 tablespoons sliced shallots

½ tablespoon minced garlic

1 tablespoon chopped cilantro

4 sprigs cilantro

¼ pound butter

1 tablespoon olive oil

RECIPE

Place the olive oil in a stainless steel pot and sweat down shallots and garlic without any caramelization. Add the vodka and Meyer lemon juice, and reduce by two thirds. Add butter and whisk to emulsify.

Next add cracked pepper and the chopped cilantro to the sauce.

Place the chucked oysters in a 450° [230°C] oven for 4 minutes or until warmed through.

Place the sauce on top of the oysters, garnish with cilantro sprigs, and serve.

Serves 4

Recipe by Rabah Abusbaltan, Chef de Cuisine, One Market Restaurant, San Francisco, and courtesy Skyy Spirits

But while Gorbachev's anti-alcohol sentiment was well intentioned, the campaign itself was perhaps shortsighted. According to Nicholas Faith and Ian Wisniewski, authors of

Classic Vodka, these policies failed because they were "applied in old-style Communist fashion, imposed from top down, and did not attack the roots of the problem, which were the country's social and economic conditions." In the late 1980s, the cash-strapped Soviet government started easing alcohol restrictions, and by 1993 all reductions in consumption had been reversed.

Today, Russia remains enamored with vodka. Russians consume more than 250 million cases of vodka per year (more than seven times the amount consumed in of the United States), with the average Russian consuming around 4 gallons (15l) per year, though exact figures aren't available. At the same time, there's been a flood of inexpensive and boot-legged vodkas in Russia, with more than 400 domestic brands now on the market. As a result, revenue from alcohol accounts for about 2 percent of the state's income, a fraction of what it was in Soviet times.

For the American and European premium brands enjoying success in the Western world, you'd think that Russia would be a burgeoning market. But for most companies it's been nothing but a bad hangover. Nationalistic pride makes competing with Russian brands difficult, while the complications of a crumbling infrastructure and inefficient systems of distribution create additional problems. But mostly it's import duties that have kept the swanky Western brands from flying off Russian shelves. "Taxes make it very difficult," says vodka

"Not a single deal, not a single

diplomatic agreement will be successful

unless it is toasted in vodka."

Russian Foreign Minister Andrei Kozyrev, 1995

OPPOSITE: *The Ice Hotel in Jukkasjärvi, Sweden, is rebuilt every year entirely of ice. The drink of choice at the Ice Hotel bar? Vodka, of course.*

marketer Michel Roux. "Absolut did well in Russia until the tax increased the price over two and a half times. It's impossible for the average Russian to pay for that."

At present Russia is also experiencing a massive proliferation of illegal alcohol production. By some accounts, boot-legged brands make up 70 percent of domestic sales. Bottles of illicit vodka sell for as little as $1.50 per liter. The effect on the nation's health has been nothing less than devastating, a result of both heavy consumption and potentially poisonous products. Estimates on the number of deaths caused by tainted vodka run from 35,000 to 100,000 per year, depending on whom you ask. One recent incident in the Siberian city of Krasnoyarsk resulted in the poisoning of 22 people, and the confiscation of 300,000 bottles of bootleg vodka. It is said that authorities found the seized stock so vile, they carted it off to be processed into brake fluid.

Combine a culture of massive alcohol consumption and a universal lack of health-consciousness with a proliferation of bootlegged brands without state quality assurances, and the result is a serious health crisis. According to research being performed at Harvard University, male life expectancy in Russia now stands at 57 years, and the country's death rate has increased by 40 percent from 1990 to 1994. These figures are not just alarming, they're unprecedented. Researcher Elizabeth Brainerd found that "declines in life expectancy of this magnitude in only four years are unparalleled in the twentieth century among countries at peace and in the absence of major famines or epidemics."

While the exact nature of Russia's vodka problems are sometimes difficult to grasp, and even harder to solve, a glimpse to the west may prove helpful. The United States has come a long way in grappling with its own consumption

patterns, and the last decade has been encouraging. The United States emerged from World War II into a zipper-mouthed, macho era that kept its problems in the closet. From there the country slid into the "anything goes" mentality of the sixties and seventies, where excess ruled. Today, Americans have achieved a better understanding of alcohol's place in society, an understanding gained through open dialogue and a renewed appreciation for quality. Granted, the United States is no problem-free zone. Many difficulties, such as the ills of teenage drinking and drunk driving—as well as alcoholism—still affect American society greatly. But there are signs of progress.

The complex problems surrounding Russian attitudes toward alcohol and drunkenness could fill a volume. Perhaps aspects of America's upscale moderation contain some answers. We'll leave the last word on this subject to vodka historian William Pokhlebkin, who advocates a "wide assortment of high-quality vodkas" in a society where "drunkards requesting treatment should be supplied with every facility to overcome their dependency." His ultimate dream? "It would be good to think that some day there will be a generation of children who have never seen a drunken brawl or a wretched alcoholic." Amen, brother.

FLAVORED AND INFUSED VODKAS: A NEW IDEA THAT'S NOT SO NEW

The folks who make and market vodka have always done a good job of learning from the past. From the revival of ancient recipes to the return of the copper pot still, many distillers are taking cues from their predecessors when developing new vodka varieties. Nothing symbolizes this reverence for the past more than today's fascination with flavored and infused vodkas.

For a unique vodka experience, try Miami's Red Square Cafe, where more than one hundred frozen vodkas from around the globe are served in a unique, Russian-inspired atmosphere. In the kitchen, chef Robbin Haas (above) serves up an eclectic menu of Russian, French, Continental, and American cuisines, including a succulent stuffed lobster (left).

COSMOPOLITIAN

INGREDIENTS

2 parts vodka

2 parts cranberry juice

1 part lime juice

¾ part Cointreau

RECIPE

Pour over ice, shake or stir, strain, and serve in a martini glass.

These days, the vodka shelves in some liquor stores can look a lot like an ice-cream parlor. The most time-honored flavors are well represented, such as citrus, berries, and peppers. Credit Swedish powerhouse brand Absolut for paving the way with their Absolut Citron, one of the top-selling vodkas in the United States. The company followed soon afterward with the popular Absolut Kurant.

These days more exotic flavors are starting to pop up. Stolichnaya is having tremendous success with their Stoli Vanil (vanilla) and Stoli Razberi (raspberry) varieties. And who would have thought that anyone in Finland had even tasted a tropical fruit? But sure enough, that country's flagship brand is making a mark with Finlandia Pineapple. But as popular as flavored vodkas are becoming, this is one trend that's back in vogue for very different reasons than had existed years before.

Flavoring vodkas with fruits, vegetables, herbs, or spices is not a new idea. The earliest distilled spirits were almost always laced with some sort of additive. In his *A History of Vodka*, historian Pokhlebkin describes how in early Russian distilling "hops and herbs—known collectively as *zel'e*—were used to give the vodka added aroma and, it was then believed, strength." The flavoring trend that abounds today has much to do with marketing, but it's safe to say that the distillers of the Middle Ages had more important things on their minds than shelf space and brand extensions. For the most part, they were simply trying to make their product fit for consumption.

Early distillation was not high art. The technology to produce what we know today as "clear vodka" wasn't commonplace until the nineteenth century, when the use of charcoal filtering became prevalent. Before that, the keepers of the stills often relied on a pinch of herbs or a slice of fruit to help their medicines go down. And let's not use the word "medicine" lightly. Many early spirits were produced for the pharmacy rather than the cocktail lounge, and the addition of honey or spices eased the taking of these remedies centuries before Mary Poppins was pushing a "spoonful of sugar."

Credit vodka lover Peter the Great for being a pioneer in flavoring vodka. Peter's love of architecture and culture was exceeded only by his love for vodka, which he enjoyed liberally. Very liberally. So it was no surprise that after a sixteen-month jaunt through western Europe in 1697–98, Peter would combine his national drink with the new flavors he discovered in Germany, Austria, and England. In addition to bringing back foods and fashions, Peter also brought a taste for pepper, berries, and spices. One of his favorite tricks was to mix vodka with anise water, and then distill it again. While his new concoctions were a hit among the socialites of St. Petersburg, the spectacular city that was his namesake, it's doubtful that Peter the Great would comprehend his influence on the tastemakers of the late twentieth century.

OPPOSITE: *Infused vodka drinks are the specialty of the house at Infusion Bar & Restaurant, a favorite of the technology industry's elite in San Francisco's SOMA, or "south of Market area."*

DEVILISH MUSSELS

INGREDIENTS

2 pounds mussels, cleaned

1 green bell pepper, chopped

1 large white onion, chopped

3 cloves garlic, minced

3 or 4 tomatoes peeled, seeded, and chopped

3 cups chicken stock

1 cup celery, chopped

1 cup chopped parsley

½ cup green onions, chopped

¼ cup fresh basil, chopped

2–3 oz. Inferno Pepper Pot Vodka

Salt and pepper, to taste

Tabasco

RECIPE

In a large pot, sauté garlic and onions over medium heat, until onions are transparent. Add celery, bell pepper, and tomatoes and cook 10 minutes over medium heat.

Add chicken stock and vodka, stirring well.

Add parsley, basil, green onions, salt, pepper, and Tabasco.

Bring to a slow boil, and add mussels to the pot. Cook over medium heat until the shells of the mussels open competely, indicating that they are done.

Remove from heat, and allow mussels to stand for 10 minutes to absorb all the flavors.

Serve over rice or pasta.

Serves 2 to 3

Recipe courtesy of Kittling Ridge Estate Wines & Spirits, makers of Inferno Pepper Pot Vodka

of cumin, fennel, anise, and honey. But the boldest souls in Sweden favor an aquavit made from wormwood, the bitter, aromatic plant best known as the main ingredient in absinthe.

For the growing league of vodka fans that want the ultimate in flavor in their vodka, the freshest approach is infusion. Now that the landscape is filled with superpremium clear vodkas, many bars and restaurants, as well as consumers, have taken it upon themselves to mix fruits, herbs, and vegetables with vodkas. Although the resulting tastes can seem quite complex, the process of making an infusion can be very straightforward. "It's embarrassingly simple," says Eric Sothern, general manager of Infusion Bar & Restaurant, a San Francisco establishment specializing in infused vodkas. The recipe? Take a large glass decanter, fill it with fresh fruits or vegetables, add vodka, and wait one to three weeks. If you've done your chemistry right, you'll have a delicious infused vodka.

Among today's proliferation of vodka brands, flavored varieties have as much to do with marketing as they do with taste. Since it can't be aged like scotch or brandy, vodka is "limited in doing something that gives alternatives to customers," says Michel Roux. "Flavored vodka is a nice extension of the brand that appeals to younger consumers," he goes on, calling it a big part of the future of vodka. Yet in this new superpremium era, it's not surprising to know that the flavored vodkas that do well are the ones that first concentrate on a quality clear product. "You can't throw flavors around," reports Magnus Philipson of Znaps Vodka, a popular brand. "You can only market flavors once you have a stable vodka."

OPPOSITE: *Inferno Vodka includes a "911" pepper in every bottle. This Canadian "pepper pot" vodka proves too hot to handle for Inferno CEO John K. Hall, who proves that things aren't all that chilly in the Great White North.* **PAGE 118:** *Antique vodka bottles from the Levize collection.* **PAGE 119:** *Pressure gauges keep the vodka under control at the Inferno distillery.*

CHAPTER 5

VODKA PLANET:
A WORLD TOUR

TODAY'S WORLD OF VODKA IS QUITE EXPAN-SIVE, AND ANY VOYAGER LOOKING TO TRAVERSE IT BETTER BE PREPARED TO LOG SOME MILES. THERE WAS A TIME WHEN THE SEARCH FOR PREMIUM VODKA WOULD LIMIT YOUR TRIP TO SWEDEN, POLAND, AND RUSSIA. IF YOU WERE LEAVING TODAY, THERE'S NO DOUBT THAT THESE DESTINATIONS WOULD BE FIRST IN YOUR TRAVEL PLANS. BUT VODKA IS NOW TRULY INTERNATIONAL.

CANADIAN ICEBERG—CANADA

If you're from the school that says exotic waters make the vodka, this is your stuff. Canadian Iceberg Vodka uses—you guessed it—12,000-year-old icebergs as the source of water for its triple-distilled natural grain spirit made from Ontario sweet corn.

CANE JUICE VODKA—BELIZE

Who said that only rum is made in the Caribbean? Billing itself as the first "tropical vodka," Cane Juice Vodka is made from 100 percent Belize sugarcane and is filtered by agitating powdered charcoal with the distilled spirits. They've always done things a little differently down there, haven't they?

CARDINAL—HOLLAND

Named after the bird that was a universal symbol for vodka in the early part of the twentieth century, this Dutch vodka is triple distilled from sugar beets in pot stills.

CHOPIN—POLAND

Along with Belvedere, Chopin is one of the two superpremium Polish vodkas becoming a hit in the United States. This potato vodka gets its name from the celebrated Polish composer.

OPPOSITE: *In the world of vodka, the wall has surely fallen. Amidst the new world order, the globe's finest vodkas make their way to the tables of Pravda in New York City.* **PAGE 126:** *A voracious appetite for vodka and a proliferation of bootlegged brands have resulted in a serious health crisis in Russia. Perhaps aspects of America's upscale moderation contain answers?*
PAGE 127: *North of GUM department store in Moscow, the bright details of a church doorway reveal the roots of the decor in countless modern, Russian-inspired cafes.*

CRATER LAKE VODKA—UNITED STATES

From the land of microbrewed beer comes Oregon's Crater Lake Vodka. This spirit is handcrafted using crystal-clear Cascade Mountain water and is filtered ten times through charcoal and Oregon volcanic rock.

CRISTALL—RUSSIA

Cristall is one of Russia's best-known distilleries, and its premium quality is known throughout the world. Made under supersecret conditions—"softening the water by reverse osmosis"—Cristall is the epitome of Russian vodka.

EVERCLEAR—UNITED STATES

Everclear is virtually an American household name due to its .infamous 180 proof grain alcohol, though now it produces a 100 proof grain spirit for us mere mortals.

FINLANDIA—FINLAND

This venerable old brand was one of the first premium vodkas in the United States when it made its appearance there in 1970. Produced from barley grain and glacial spring water, Finlandia is now the third leading imported vodka in the United States (behind Absolut and Stolichnaya). Not to be left out of the flavored vodka craze, Finlandia produces Arctic Cranberry and, in a true case of culture clash, Arctic Pineapple.

FRIS VODKA SKANDIA—DENMARK

Fris (from the Danish words meaning "frost" and "ice") is a 100 percent grain neutral spirit that's been available in the United States since 1992. Before it undergoes six distillations and three filtrations, the water used to produce Fris is chemically altered to lower the freezing point of water, which is said to "alter the texture" of the vodka.

GORILKA—UKRAINE

This vodka was originally brought to Moscow by Cherkassk Cossacks. Today Gorilka is a grain spirit flavored with natural linden.

VODKA-MARINATED SALMON

INGREDIENTS

1 pound salmon

1 ounce vodka

⅓ ounce chopped juniper berries

½ teaspoon saffron

¾ ounce chopped dill

⅔ cup granulated sugar

⅓ cup sea salt

RECIPE

Fillet the salmon, leaving the skin on. Sprinkle vodka, juniper berries, saffron, and dill over the flesh of the salmon. Make a mix of the salt and sugar; sprinkle some over the salmon. Place the fish in a nonreactive container, flesh side up, and sprinkle with the remainder of the salt and sugar mix. Fill the container with water to cover the salmon, cover with plastic, and allow to marinate in a warm place overnight.

Serve with black caviar and crème fraîche.

Serves 4

Recipe courtesy of Klaus Kochendoerfer, Executive Chef, Grand Hotel Europe, St. Petersburg

OPPOSITE: *Klaus Kochendoerfer's vodka-marinated salmon, served with caviar, 'the' best of Russian food and drink.* **PAGE 130:** *A female worker at the Irkutsk distillery in Siberia.* **PAGE 131:** *A holding tank at the Levize distillery in St. Petersburg, Russia.*

GREY GOOSE—FRANCE

From the land of bubbly comes Grey Goose, produced in Cognac, France. Made from spring water and filtered through champagne limestone, the French vodka in the frosted bottle is popping up on top shelves all across the United States.

INFERNO PEPPER POT VODKA—CANADA

From the icy region of Ontario, Canada, comes Inferno, one of the hottest vodkas around. Quadruple distilled and cold charcoal filtered, Inferno vodka includes a fiery "911" pepper in every bottle—perfect for warming up winter nights.

KETEL ONE—HOLLAND

One the most talked-about new brands in American, Ketel One is anything but new. The Nolet distillery in the Dutch port city of Schiedam has been distilling spirits since 1691. The family was so tight with the Russian tsar that after the 1917 revolution, they were granted the right to use the double eagle from the tsar's own crest. Ketel One was introduced in the United States in 1990, rising to the number four imported vodka by 1996. Ketel One is made from wheat grain and is distilled the old-fashioned way in pot stills, where only the middle of a batch (the "heart" of the distillate) is used.

LUKSUSOWA—POLAND

Few vodkas are more associated with Poland than Luksusowa. Named after the Polish word for "deluxe," Luksusowa is triple distilled from potatoes at the historic Lancut Distillery, located in the southeastern region of Poland on the fringe of the Carpathian foothills.

MOSKOVSKAYA—RUSSIA

A classic rye Russian grain spirit, straight from the motherland, Moskovskaya (named for the Russian capital) also comes in a lemon variety, as well as in a superpremium variety named Cristall for the famous distillery where it's made.

TEXAS VODKA:
Ultrapremium Oxymoron

If you're looking to tour the world of vodka, chances are Texas isn't on your itinerary. That's not surprising. Mention vodka, and most people's minds go decidedly frigid, conjuring images of such arctic locations as Russia, Poland, or Sweden. Not that the land of the Rio Grande isn't drink friendly. The lone star state ranks second in beer production and ninth in wine production in the United States, but distilling has historically eluded the Texas border. Until now.

It took a lone cowboy by the name of Tito to bring distilling to Texas, in 1996. After a long wrangle with state and federal authorities, Tito opened Fifth Generations Inc. "We're Texas's first and only distillery," says owner Tito Beveridge (honest, that's his name).

To say that Tito's Handmade Vodka is indeed handmade is an understatement. Fifth Generation is virtually a one-man shop, and the grain spirit rolls out of the stills in small 175-gallon (750l) batches. Tito's original still was crafted out of a sixteen-gallon (68l) beer keg and, even today, Tito's old-fashioned pot still is a funky conglomeration of duct tape and used restaurant equipment.

Only a Texan would have the tenacity to get a distillery up and running in a land that has no history of distillation (the state once had a rectifier, but it closed in 1988). From the beginning, Tito was determined to form a legit business, and so started a long and weird odyssey into the bowels of government regulation. "The TABC (Texas Alcoholic Beverage Commission) had never dealt with a distillery before," Tito says. "You're kind of on their time clock." After being subject to the whim of regulators, Tito found himself in the vodka equivalent of the Mexican standoff: you have to be a functioning distillery in order to get a permit, and you can't be a functioning distillery without a permit. "You're supposed to be a distillery without having distilled," Tito says with a sigh. After many months of sorting through federal codes and producing vodka he couldn't sell, Tito finally roped his license.

By now, Tito has let go of the startup headaches and relishes producing a quality vodka. "I make a microdistilled, handcrafted product," he says of his clear grain spirit, distilled six times (twice in a pot still). More than a way of staying out of jail, Tito sees his hard-fought legitimacy as being essential to making a good vodka. "When you're on the edge of the law, some guys might have pride in their juice, but most are just trying to get the alcohol made," he says in his long Texas drawl. "I pride myself on making a quality spirit."

OKHOTNICHYA—RUSSIA

Russian flavored vodkas like Okhotnichya, popularly known as "hunter's vodka," were originally used as tip-backs after aristocratic hunts. Okhotnichya's long line of ingredients include ginger, cloves, red and black peppers, juniper, coffee, anise, and orange peels. The spicy flavors are said to warm body and soul after a cold day in the great outdoors.

OPPOSITE: *Vodka cocktails by candlelight at the Red Square Cafe in Miami.*

PERLOVA—UKRAINE

Perlova is 100 percent wheat vodka made from a centuries-old recipe using "soft" water from the Carpathian Mountains region and filtered through charcoal and quartz crystal.

VENISON MARINATED IN VODKA

INGREDIENTS

Two ⅓-pound venison fillets

⅔ ounce vodka

1 ounce lingonberry jam

½ ounce lingonberries

3 tablespoons vegetable oil

2 ounces criminis or other wild mushrooms

2 ounces chanterelles

1 small onion

½ cup venison gravy (recipe follows)

Chopped parsley for garnish

Salt and pepper to taste

RECIPE

Marinate the venison fillets overnight in the vodka, lingonberries, and jam. In a medium saucepan, sauté the onion in 1 tablespoon of the vegetable oil until translucent; then add the mushrooms and cook until the juices have been released and the mushrooms are tender. In a separate saucepan, brown the venison fillets in the remaining oil until medium rare, about five minutes.

Let the meat rest for two minutes, and then cut into slices and serve with the mushrooms and hot gravy.

Garnish with parsley, and add salt and pepper to taste. Serve with mashed potatoes.

VENISON GRAVY

INGREDIENTS

2 cups beef stock

½ tablespoon fresh thyme

⅓ cup finely diced carrots

⅓ cup leeks, sliced in thin half-moons (cut in half lengthwise and wash thoroughly in cold water to remove sand)

⅓ cup finely diced celery

2 tablespoons unsalted butter

1 tablespoon vegetable oil

Salt and pepper to taste

Melt butter over medium heat and add oil. Sauté vegetables and thyme in butter mixture until soft, about 10 minutes. Add stock and simmer over low heat until liquid has reduced by half, about 20 minutes. Strain, reserving vegetables, and return sauce to pan to keep warm. If it seems too thin, puree cooked vegetables in just enough additional stock to allow blender to work, strain again, and add a tablespoon or so of puree to sauce. Season with salt and pepper to taste.

Serves 2

Recipe courtesy of Klaus Kochendoerfer, Executive Chef, Grand Hotel Europe, St. Petersburg

OPPOSITE: *Fresh venison marinated in vodka is served at St. Petersburg's Grand Hotel Europe.*

ABSOLUT BEEF WOK

INGREDIENTS

1 pound beef

2 large onions, sliced

3 ounces capers

2 cups seasonal vegetables, such as

Sweet peas, whole

Carrots, sliced

Red bell peppers, sliced

Broccoli, sectioned

Cauliflower, sectioned

Zucchini, sliced

String beans, whole

Marinade

3 ounces Absolut Vodka

2 ounces peanut oil

Wok Sauce

10 ounces tomato sauce

3½ ounces sweet soy sauce

2 cloves garlic, pressed

1 teaspoon fresh ginger, finely chopped

Salt and pepper

3½ ounces hot sauce or strong chili sauce

RECIPE

Marinate the beef for 3 to 4 hours.

Heat wok until very hot. Add oil, brown beef lightly, then remove.

Clean wok and add new oil. Stir-fry the vegetables. Add sauce and beef. Heat and serve with the wok sauce.

Serves 4

Recipe courtesy of The Absolut Company

PRIVIET—RUSSIA

New to the United States market, courtesy of Carillon Importers (who market Stolichnaya), Priviet is made from select winter wheat and glacial water.

RAIN—UNITED STATES

If there were an award for the most politically correct vodka, Rain would be a contender. Rain Vodka is made from organically grown Illinois grain and Kentucky limestone water. Rain's cartons and labels are produced from recycled paper, and the company makes a contribution to the Wilderness Society for each bottle sold. Talk about a no-guilt drink!

RAINBOW—UNITED STATES

Rainbow Vodka is one of five "community spirits" produced by Marie Brizard Wines & Spirits. This triple-distilled grain vodka, made in Bardstown, Kentucky, strives to benefit more than just the premium vodka drinker enjoying their products. Half of the net profits from the sale of Rainbow Vodka go to local, nonprofit AIDS organizations located in the states where Rainbow is sold.

SKYY—UNITED STATES

Skyy came from out of nowhere to become one of the most popular premium vodkas in the United States. The makers of Skyy fanatically strive for a vodka that is 100 percent pure, free from all impurities (known as congeners). Made from American grain, Skyy is quadruple distilled and put through a proprietary three-step filtration system, yielding a vodka as clean as its name.

SMIRNOFF—UNITED STATES AND RUSSIA

The grand old name in vodka, Smirnoff is the best-selling vodka in the United States (and the number two spirit overall). For many years the company has marketed its standard "Red Label" Smirnoff, a grain spirit produced in the States. With the rising popularity of premium vodkas, the company has intro-

duced Smirnoff Black, a superpremium Russian variety produced in small batches in pot stills.

STOLICHNAYA—RUSSIA

You'd be hard-pressed to find anyone who isn't familiar with this classic Russian vodka and its familiar label featuring the Hotel Moscow, built by Stalin in 1932 (Stolichnaya means "for the capital"). Made from glacial water and winter wheat, "Stoli" has achieved a permanent place among premium vodkas, and is the second leading imported vodka brand in the United States. The company also markets six flavored varieties, as well as Stolichnaya Gold, a superpremium vodka.

ABOVE: Filtration is the key to a vodka's quality, and techniques are closely guarded secrets in the industry. These are the filtration tanks at Znaps distillery in England.

TANQUERAY STERLING—UNITED KINGDOM

Tanqueray, best known as a venerable gin, is also a permanent resident on the vodka side of the top shelf. Made from Scottish water and filtered through granite chips, Tanqueray Sterling Vodka is now the fifth leading imported vodka in the United States.

TETON GLACIER POTATO—UNITED STATES

If you're going to make an American potato vodka, where would the spuds come from? Idaho, of course. Throw in some pure Rocky Mountain well water and you have Teton Glacier Potato Vodka, made in Rigby, Idaho.

TITO'S HANDMADE VODKA—UNITED STATES

If you're looking for vodka pioneers, look no farther than Texas. Tito's Handmade Vodka is the product of the lone star state's first and only distillery. Lovingly produced in homemade pot stills, Tito's is made from American corn and is distilled six times.

OPPOSITE: Filtration tanks at Levize distillery in Russia.
ABOVE: Russian vodka makers often attribute the unique character of their vodkas to the icy lakes from which they draw water. PAGE 140: *On the job at a Russian distillery, these women are putting the finishing touches on a batch.*
PAGE 141: *The final product, packed in ice and ready for tasting at New York's Pravda.*

ULTRAA—RUSSIA

Ultraa prides itself on being Russia's "modern" vodka. The spirit is made from the "oxygen saturated" water of Russia's Lake Lagodas and filtered through quartz sand.

VIRGIN—UNITED KINGDOM

A popular brand in England, Virgin is a triple-distilled spirit courtesy of whiskey maker William Grant & Sons and the Virgin Trading Company, another venture of the ubiquitous rock and roll/airline company.

WOLFSCHMIDT—UNITED STATES

An American brand, Wolfschmidt claims royal roots. The Wolfschmidt family was said to have been the distiller to Russian tsars. Today, Wolfschmidt is a grain spirit made in the United States.

WODKA WYBOROWA—POLAND

Produced at the historic Lancut Distillery, Wyborowa is made from premium rye grain spirits and put through a "painstaking rectification process." Wyborowa is one of the best-known Polish vodkas, and is marketed in two strengths and five different flavors.

ZNAPS—ENGLAND

Truly an international brand, Znaps is made from Swedish raw materials distilled in Birmingham, England at Manor Brewery. Znaps is available as a clear spirit and in seven flavored varieties.

ZUBROWKA—POLAND

Zubrowka is also known as "bison vodka," owing to the fact that it gains its character from the addition of bison grass in the distillation process. Zubrowka's greenish yellow color and grassy aroma make it one of the world's most unusual vodkas.

APPROXIMATE METRIC EQUIVALENTS

1/4 teaspoon	=	1.23ml
1/2 teaspoon	=	2.46ml
3/4 teaspoon	=	3.7ml
1 teaspoon	=	4.93ml
1 tablespoon	=	14.79ml
1/4 cup	=	59.15ml
1/3 cup	=	78.86ml
1/2 cup	=	118.3ml
1 cup	=	236.59ml
1 pint	=	473.18ml
1 quart	=	946.36ml

SELECTED BIBLIOGRAPHY

Brown, Gordon. *Classic Spirits of the World.* New York/London/Paris: Abbeville Press, 1995.

Conrad, Barnaby III. *The Martini.* San Francisco: Chronicle Books, 1995.

Embury, David A. *The Fine Art of Mixing Drinks.* Garden City, NY: Dolphin Books, 1961.

Faith, Nicholas, and Ian Wisniewski. *Classic Vodka.* London: Prion Books, 1997.

Grimes, William. *Straight Up or On the Rocks: A Cultural History of American Drink.* New York: Simon & Schuster, 1981.

Jeffers, H. Paul. *High Spirits: A Celebration of Scotch, Bourbon, Cognac, and More...* New York: Lyons & Burford, 1997.

Lanza, Joseph. *The Cocktail: The Influence of Spirits on the American Psyche.* New York: Picador USA, 1995.

Lipinski, Robert A., and Kathleen A. Lipinski. *Professional Guide to Alcoholic Beverages.* New York: Van Nostrand Reinhold, 1989.

Miller, Anistatia R., and Jared M. Brown. *Stirred Not Shaken: A Celebration of the Martini.* New York: Harper Perennial, 1997.

Pokhlebkin, William. *A History of Vodka.* Translated by Renfrey Clarke. London/New York: Verso, 1992.

Regan, Gary, and Mardee Haidin Regan. *The Martini Companion: A Connoisseur's Guide.* Philadelphia: Running Press, 1997.

INDEX